The Resourceful Physics Teacher

The Resourceful Physics Teacher

600 ideas for creative teaching

Keith Gibbs

Queens's College, Taunton

Institute of Physics Publishing
Bristol and Philadelphia

British Library Cataloguing-in-Publication Data

A catalogue record for this book is available from the British Library.

ISBN 0 7503 0581 9

Library of Congress Cataloging-in-Publication Data are available.

Reprinted 2000

Commissioning Editor: Gillian Lindsey
Editorial Assistant: Victoria Le Billon
Production Editor: Adrian Corrigan
Production Control: Sarah Plenty and Jenny Troyano
Cover Design: Kevin Lowry
Marketing Executive: Colin Fenton

Published by Institute of Physics Publishing, wholly owned by The Institute of Physics, London

Institute of Physics Publishing, Dirac House, Temple Back, Bristol BS1 6BE, UK

US Office: Institute of Physics Publishing, The Public Ledger Building, Suite 1035, 150 South Independence Mall West, Philadelphia, PA 19106, USA

Typeset in the UK by Mackreth Media Services, Hemel Hempstead, Herts

Printed in the UK by J W Arrowsmith Ltd, Bristol

CONTENTS

INTRODUCTION

A little over fifteen years ago, when I submitted part of the manuscript of my first book to a publisher, it was returned with the comment that it was 'merely a collection of oddments'. Well, all these years later, perhaps that is what this new venture is all about!

My first aim on starting to write this book was to bring together a collection of interesting demonstration experiments and ideas that I have come across in thirty years of physics teaching in a number of schools. I do not claim that all the ideas are my own; indeed, many have been suggested by relatives, friends and colleagues, not to mention many of my students, past and present, who have encouraged me over the years, and many others have been gleaned from magazines, lectures and other publications. I am most grateful to all those who have given their expertise. Since the initial period of writing I have added more explanation and background theory to help those teachers whose basic specialization is not in physics.

I have tried to make the book of use to all physics teachers, both new and experienced. I am sure that many of those who read it will come across old favourites but I hope that all readers will find at least something that is new and challenging.

I like analogies and I have included a number of those within the text. We all have our own favourites but I hope that teachers will be able to use some of mine or at least get some ideas for alternative ways of explaining things.

I also hope that the book will serve to popularize the subject and make people realize that there is much of fun and interest in physics. At a dinner party once, on finding out what my job was, a lady said to me 'Ah, physics – end of conversation!'. I really hope that these pages will show that physics does have something to offer.

Safety considerations must always be borne in mind when demonstrating experiments or suggesting ones for students to carry out themselves. However, if a zero-risk policy were to be adopted, we would do no practical work at all!

I would like to thank staff at Institute of Physics Publishing in Bristol for their enthusiasm and support for the project, especially Gillian Lindsey. I must also recognize the help given to me by Lucy Broadley and members of my sixth form physics group in checking the text; however, any remaining errors must be my responsibility.

Finally, I hope that those of you who read and use this book will find it helpful, interesting and enjoyable!

Keith Gibbs
Taunton 1999

SAFETY CONSIDERATIONS

By their nature, some experiments in this book are inherently dangerous. Specific hazards are identified within the text by the use of the following general warning sign:

The hazards highlighted are not necessarily exhaustive and teachers must use extreme caution at all times when carrying out experiments. Teachers must always follow good practice, including that given listed in the publications below.

Good practice

Safety glasses should be worn whenever there is the slightest risk to the teacher's or pupils' eyes; a safety screen should be placed between the teacher and the audience if there is the remotest risk of even a tiny explosion. A heatproof mat should be placed on the bench if any experiments involving heating are to be performed. A large tray should be placed under any apparatus where a corrosive or staining liquid might be spilled.

Suggested publications

Topics in Safety 2nd edition, Association for Science Education, 1988
Safeguards in the School Laboratory 10th edition, Association for Science Education, 1996

GENERAL

This first section deals with some general suggestions and ideas that could cover the whole area of physics teaching. I begin with a couple of quotations about physics.

1. Quotations
2. How odd are you?
3. Estimates of quantities
4. Dimensional farm animals
5. Gradients
6. Use of a closed-circuit TV camera
7. Romance and physics?
8. Simple measurements in physics
9. Getting values for the sizes of small quantities

1 Quotations

'Physics is a subject that if you understand it, you do not have to know much.' *Susan Kalmus, a former member of one of my Lower Sixth classes.*

'Ah, physics . . . end of conversation!' — *A lady at a dinner party.*

2 How odd are you?

This set of measurements is a useful introduction to physics or general science for students as they arrive at a new school. It gets them to collect and present results, draw block graphs, work out averages and also to get to know the other members of the class.

Ask them to make the following measurements of the other members of the class:

(a) height
(b) hand span
(c) pulse rate (resting)
(d) eye colour.

Try to present them in a form that means something! Using the heights of the class to make a block graph is best for this analytical part of the experiment. The maximum height, the most common height range and the average height of the class can then be found.

3 Estimates of quantities

It is always important in physics to have a rough idea of the values of quantities that you are attempting to measure. Introduce this idea by a set of intelligent guesses. Here are some suggestions.

Get the students to guess:

(a) the temperature of some hot water (at about 50°C) in a bucket by putting their hands in it;

(b) the length of a time interval of twenty seconds by closing their eyes and counting while a friend measures the interval with a stop clock;

(c) the mass of a lump of stone (say 10 kg) or a building block by lifting it up;

(d) the length of the lab just by looking at it — no cheating by pacing it out.

This is not always so easy; I remember standing above Bryce Canyon in Utah and being asked by the guide how far away I thought a mountain on the horizon was. The air was beautifully clear and I guessed thirty miles — it was actually a hundred miles away! The Apollo astronauts also had great difficulty estimating distances on the Moon. Owing to the lack of atmosphere distant mountains were not blurred, leading to problems with perspective and hence navigation.

Apparatus required: •Bucket of hot (50°C) water •Stop clock •Thermometer •Measuring tape •Large rock (10 kg).

4 Dimensional farm animals

This demonstration is designed to show the difference between quantities with different units and dimensions. I use a collection of toy farm animals: three pigs, three sheep and three cows.

Take one or two pigs, a couple of sheep and a cow and ask 'How many are there?' The students should then reply with the question 'How many what?'

Can you add up cows and pigs? Not really. The whole point is that it is still just a mixture of animals. In the same way you cannot add metres and kilograms. I find it an amusing way of introducing students to the higher level course.

Thinking of equations use a few of the animals to show that you must have the same 'power' of animals on each side of an equation. Is a pig on its back '$(pig)^{-1}$'?

Apparatus required: •Set of farm animals.

5 Gradients

Accurate drawing of a gradient to a curve can be made simpler by using a plane mirror. This should first be placed standing on your graph and at right angles to the line. It is relatively simple to adjust it until the curve and its reflection join smoothly without a kink at the reflecting surface. Now draw along the back of the mirror to give a normal (90° line) to the curve. Finally draw a line at right angles to this to get the gradient of the curve at that point.

6 Use of a closed-circuit TV camera

After teaching for many years I have at last bought a small TV camera. This has made an enormous difference to some demonstration experiments. You can use either a camcorder or a security camera (this cost the school about £300 for a colour model). The latter type is very small, can be easily mounted on a retort stand base and gives excellent quality. I have used it for the following purposes.

(a)   To make small-scale demonstrations visible to the whole class. If you think the students can see everything, go to the back of a middle sized class and look back at the front desk for yourself!

(b)   In place of a vernier callipers as a measuring device by putting a ruler graduated in half millimetres in the field of view and then taking all measurements of both the scale and the object from the TV screen.

(c)   To produce a large image of a cathode ray oscilloscope screen on the TV; the waveforms can then be seen by the whole class.

(d)   To make recordings of tricky experiments for later use.

(e)   To observe infrared radiation; see the section on radiation.

7 Romance and physics?

Imagine a romantic evening; how could physics explain the effects?

How beautiful your eyes are could become: I see the molecules in your iris have not joined together to form longer chains and so you have blue eyes.

That is a beautiful sparkling diamond ring: the refractive index of that transparent material in your ring is large and it has been cut to show multiple internal reflection.

The waves on this beach are breaking gently on the sand: I see that the velocity of the circular motion of the water particles is decreasing as they move up the beach.

What a beautiful sunset: the electromagnetic radiation of longer wavelength from the nearest star has not been scattered so much and the sky to the west is red.

Look, here is a wishing well: Let's estimate its depth by dropping a penny in and seeing how long it takes to reach the bottom.

I suggest that perhaps these versions are not really what you would want to say!

---

8   Simple measurements in physics

As an introduction to measurement get the students to measure:

(a) the length of a line,
(b) the distance between two dots,
(c) the diagonal of a rectangle,
(d) the diameter of a circle.

Prepare both large and small shapes to experiment on.

---

9   Getting values for the sizes of small quantities

Stress to the students that the mass of a small ball bearing can be found by measuring the mass of a known large number. The thickness of a sheet of paper, for example is best found by measuring the thickness of a number of sheets (or by using a micrometer).

The mass of a paper pellet can be found by finding the mass and area of a sheet of paper and then finding the area of paper used to make the pellet.

PRESSURE BETWEEN SOLIDS

General theory for this section

The greater the area of contact the less the pressure between two solid surfaces.

Pressure = Force/Area Units for pressure are Pascal (Pa).

$1 \text{ Pa} = 1 \text{ Nm}^{-2}$.

1. Pressure: bed of nails
2. Pressure: penny and hook
3. Pressure: cheese and wire, rucksacks and shoulders
4. Pressure: elephant and girl
5. Student pressure
6. Standing on gravel and sand

1 Pressure: bed of nails

How can people lie on a bed of nails without excruciating pain? It is all to do with the area of contact. Although each nail has a sharp point, the total area of the ends of the nails makes the pressure on any one part of the body bearable. A demonstration of this (I do not recommend getting pupils, or staff, to lie on beds of nails) is to use a bed of nails with a loaded balloon to represent the body. Further experiments can be carried out with a weight resting on the balloon to show how much it can stand without bursting. A very careful demonstration using a compression newton meter and a drawing pin could be used to find out the pressure exerted on a thumb before it hurts!

Apparatus required: •Board with nails facing upwards •Board without nails •Balloons •Kilogram masses.

2 Pressure: penny and hook

(a) This is a very simple example of the dependence of pressure on area of contact. Put an S hook on the end of your finger and suspend a 1 kg mass from it: it hurts. Now do the same thing but have a small coin between the S hook and your finger: it hurts much less. The pressure on your finger is much lower since although the force is the same (or very nearly if you

disregard the weight of the coin) the area of contact is much greater.

Apparatus required: •S hook •1 kg mass •Small coin such as a 1p piece.

---

3   Pressure: cheese and wire, rucksacks and shoulders

These two demonstrations are good practical examples of pressure between solids.

(a) Cut through a piece of cheese with a cheese wire and actually measure the pressure on the cheese. You simply need to pull the wire down with a newton meter and use a micrometer to measure the diameter of the wire.

(b) Get one of the pupils to put a rucksack on their shoulders and then load it with weights; the straps begin to press and it hurts. Then increase the area of contact between the straps and their shoulders by putting some large pieces of foam rubber in between. It now hurts less.

Apparatus required: •Cheese •Thin wire •Micrometer •Newton meter •Rucksack •Large masses as a load.

---

4   Pressure: elephant and girl

In order to emphasize the dependence of pressure on the area of contact compare the pressure exerted by a girl in stiletto heels with an elephant standing on one foot. We usually estimate the elephant to have a mass of two or three tons (3000 kg) and a foot of area about 300 cm^2. Try and get one of the class to bring in a stiletto heeled shoe to measure!

Apparatus required: •Stiletto heeled shoe •Data about elephants •Bathroom scales.

---

5   Student pressure

The students can work out the pressure between their feet and the ground when they are standing by first finding their weight and then measuring the area of their own feet by standing on graph paper and drawing round their feet without their shoes on. A good estimate of the area of their feet in contact with the ground can be made by counting the squares. It is interesting to see who thinks that they have an instep.

Typical results for this experiment: weight of a twelve year old student 400 N; area of contact = 200 cm^2; Pressure = 2 N cm^{-2}

Apparatus required: •Bathroom scales •Graph paper.

6 Standing on gravel and sand

These two simple experiments demonstrate the effect of different areas of contact between a person and the ground.

(a) Stand on some gravel in a tray with your shoes on. Then take your shoes off and do it again; it is much more painful since the area of contact is much less.

(b) Stand on some sand in a tray with your shoes on. Now put a small block of wood on the sand and stand on that; it will sink in, showing the increased pressure although the force is unchanged. If you can get a large tray try making some hardboard shapes to fix to the underside of their shoes to increase the size of the area of contact.

The police often get an idea of the weight of burglars by the depths of the imprints of their feet in soft earth.

Apparatus required: •Tray of gravel •Tray of sand •Wood blocks of different sizes.

PRESSURE IN LIQUIDS

General theory for this section

In a liquid the pressure at a point within it depends on two things:
(a) the density of the liquid and (b) the vertical distance below the liquid surface.

Pressure in a liquid = depth (h) × liquid density (ρ) × gravitational acceleration (g).

In a liquid the pressure acts in all directions.

1. Can with holes
2. Ball with rubber tube

1   Can with holes

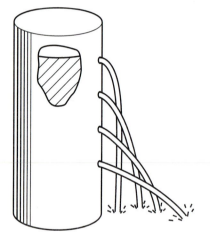

You can demonstrate the increase in water pressure by using a can or plastic bottle with a series of holes — both up and down (to show the increase in depth) and around the side (to show pressure acting in all directions). A 2m long 3 cm diameter plastic water pipe would show this much better — the large head of water will give some impressive jets near the bottom. To show this correctly small tubes should project at each orifice. A detailed consideration of Torricelli's theorem is needed to predict which jets will go the furthest.

Apparatus required: •Plastic bottle or can with holes in the side •Drain pipe with holes in the side •Water •Bucket.

2   Ball with rubber tube

Another way of showing that pressure at a point in a liquid acts equally in all directions is to take a tennis ball, or any hollow rubber ball of that size, and make a number of small holes (diameter two or three millimetres) in it. Then fix a long glass tube into the top of the ball. Pour water into the tube and show that since the pressure acts in all directions water comes out of all the holes round the ball with roughly equal force.

Apparatus required: •Ball with holes with long glass tube fitted into it •Water.

PRESSURE OF GASES AND ATMOSPHERIC PRESSURE

General theory and information for this section

The pressure of the Earth's atmosphere at sea level is about 10^5 Pa and the density of the air at sea level is 1.2 kg m^{-3}. If the atmosphere were compressed to a uniform density equal to that at the Earth's surface it would form a layer 8.6 km deep round the planet.

It is important for all these experiments to emphasize the forces **pushing** on the walls of containers or liquids rather than causing a **pulling** effect due to suction.

1. Rubber sucker
2. Lath and newspaper
3. Party blowers and air pressure
4. Balloon in bell jar
5. Pouring a can of lemonade — one hole or two?
6. Drinking through a long straw
7. Glass and card
8. Lifting a student by blowing
9. Atmospheric pressure: pump with reversed washer
10. The hosepipe problem
11. Lung pressure
12. Atmospheric pressure: the fountain
13. Bucket of chocolate blancmange
14. Collapsing can and collapsing bottle
15. Test tube in a bell jar

1 Rubber sucker

A rubber sucker gives a simple demonstration of air pressure. Press the rubber sucker against a wall or other smooth surface. It sticks to the wall since the pressure outside the sucker is greater than that inside. Moistening the edges of the rubber helps to improve the seal between it and a solid surface and prevents air leaking in.

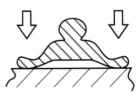

Apparatus required: •Rubber sucker or sink plunger.

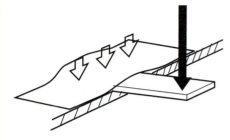

2   Lath and newspaper

This very simple experiment demonstrates the pressure of the atmosphere. Take a thin strip of wood (known as a lath) about the size of a 30 cm ruler and lay it on the table with about a third of its length projecting over the edge. Now smooth out a large sheet of newspaper so that it covers the part of the lath on the bench. Gently lift the exposed end of the lath; the paper lifts and air has had a chance to get underneath. Now smooth the paper back and hit the exposed section of the wood smartly with the edge of your hand like a karate chop; air does not have a chance to get underneath, the pressure of the atmosphere holds the paper down and the lath breaks.

Theory: pressure of the atmosphere $= 10^5$ Pa so the weight of air on a sheet of newspaper opened out to an area of 1 m $\times$ 0.75 m is 7.5×10^4 N

Apparatus required: •Thin wooden lath (20 mm $\times$ 300 mm $\times$ 2 mm). The thickness is quite critical — too thick and it won't break, too thin and it becomes too whippy and still won't break. I have also used the tongue from a piece of tongued and grooved board, which works well.

3   Party blowers and air pressure

This is a self-explanatory and simple model of a Bourdon gauge. The harder you blow, and therefore the greater the pressure inside the paper tube, the more it unwinds. The springy wire along the blower coils it up again when you stop blowing. Compare the party blower with an actual Bourdon gauge.

Apparatus required: •Party blower •Bourdon gauge.

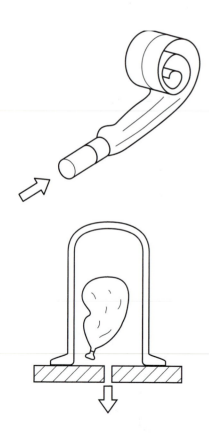

4   Balloon in bell jar

(a) Partly inflate a balloon and put it in a bell jar connected to a vacuum pump. The partly inflated balloon will inflate if the air is pumped out of the bell jar because the air pressure inside the balloon is greater than that outside. This shows what would happen to your lungs on depressurization in a plane or in space. It is also an example of what happens when a diver gets the bends. If the diver comes to the surface too rapidly so reducing the pressure on the outside of their body too quickly, bubbles of gas come out of solution in the blood.

(b) A demonstration using simpler apparatus to that just described now follows. Fix a bicycle pump valve into the side of a plastic bottle near the base. Put a balloon inside with the neck of the balloon over the neck of the bottle. Using a bicycle pump with a reversed washer, pump out some air from below the balloon; it will inflate. The bottle needs to be strong otherwise it will itself begin to collapse! A further alternative is to put a balloon in the bottle in a version of experiment 15(b) before the water is added and use a rigid bottle such as a wine demijohn.

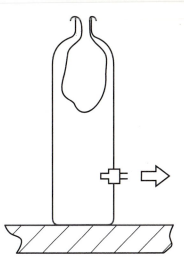

(c) Yet another variation is to use experiment (a) but with a marshmallow in the bell jar instead of the balloon. As the air pressure in the bell jar is reduced the air is drawn out of the marshmallow which should then theoretically collapse to about half its original volume! In fact the reduction in pressure causes the ones that I have used to expand — they shrink when air is let back in the bell jar. Chocolate covered ones exude a stream of white foam!

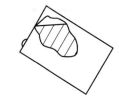

Apparatus required: •Balloon •Vacuum pump •Bell jar •Bicycle pump •Marshmallow •Plastic bottle.

5 Pouring a can of lemonade — one hole or two?

When you try and pour lemonade out of a can through one small hole you find that it is very difficult. It is worth demonstrating this by using two prepared tins, both with removable lids that can be filled with water. One has only one small hole but the other has two at opposite sides of the can. Pouring liquid out of one hole would leave a partial vacuum behind so resisting further pouring. The addition of the second hole allows air to get in.

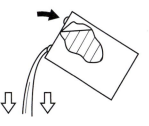

An even simpler demonstration of this is to fill a milk bottle with water and upend it over a sink — the water comes out relatively slowly in great 'lumps'. Tilting the bottle allows air to pass into it over the outgoing water and makes the pouring simpler and quicker. Emptying a wine demijohn is also made easier if it is spun to create a vortex through which air can enter when it is turned upside down.

Apparatus required: •Milk bottle •Water •Bucket or sink •Wine demijohn.

6 Drinking through a long straw

(a) Put a can of lemonade at the bottom of a flight of stairs. Get a student to stand at the top and then ask them to try and drink some of the lemonade using a long clear plastic tube. They can't, or at least they find that it is very difficult. Using a clear plastic tube is better since you can actually watch the progress of the liquid. This is a convincing demonstration of atmospheric pressure and its limitations.

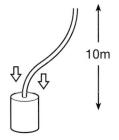

10m

Theory: The maximum height of water (or fizzy drink) that the atmosphere can support is about 10 m. Therefore, even if the pressure in the straw is reduced to zero, the liquid will rise no higher than this. In practice the height risen is much less! Height of water (drink) = Atmospheric pressure/($g \times$ density of water) = 10 m.

(b) Another way of **not** getting any drink into your mouth when you try and suck it up a straw is to put two straws into your mouth! One has its other end in the liquid while the other has its open end in air. No matter how hard you suck no liquid will rise up the tube — you are simply not able to reduce the pressure in your mouth; it remains at atmospheric pressure because one of the straws has an end open to the air.

Apparatus required: •Can of drink •Long clear plastic tube (makes the rise of the liquid visible to the rest of the class).

---

7 Glass and card

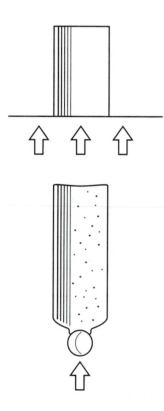

(a) Fill a glass to the brim with water; carefully slide a card over the top and then turn it upside down over a sink, holding the card in place with your hand. When the glass is upside down take your hand away. The card will stay there, it is being held onto the glass because the air pressure below the card is greater than the water pressure above it. Ask the students the following questions:

  (i) What is the longest beaker that you could invert full of water before the card comes off ?

  (ii) What happens if a little air leaks into the glass?

(b) In a variation of the previous classic experiment you can use instead a table tennis ball in the mouth of a milk bottle. If the bottle is filled with water with the table tennis ball on the end the bottle can then be turned upside down; the air pressure holds the table tennis ball in place. It also works quite well with some air in the bottle — I have done it with the bottle almost empty! (Of course there are surface tension effects; try and ignore these.)

Apparatus required: (a) •Glass •Piece of cardboard •Water •Sink or bowl (b) •Milk bottle •Table tennis ball.

---

8 Lifting a pupil by blowing

Stand one student on a wooden board which is resting on top of a plastic bag to which a rubber tube has been fitted. Blow into the bag through the tube and show that you can lift the student up!

  Pouring water down a tube into a rubber hot water bottle on which someone is sitting has the same effect.

Theory: force (= weight of the student) = pressure × area. If your lung pressure is, say, 0.1 atmospheres (10^4 Pa) above atmospheric pressure then to lift a student of weight 400 N you need a board with an area of contact of 0.04 m² (20 cm × 20 cm).

Apparatus required: Plastic bag with rubber tube sealed into the neck
Wooden board (say about 0.5 m square).

9 Atmospheric pressure — pump with reversed washer

This is a simple experiment to demonstrate the pressure of the atmosphere and can even be used to get a very rough estimate of its value. Reverse the washer in a bicycle pump and mount the pump vertically in a clamp. Hanging weights on the handle can then be used as a way of showing the pressure of the atmosphere. The weights pull downwards on the handle but the pressure of the atmosphere acting upwards on the reversed washer prevents the piston from descending.

Theory: If the cross sectional area of the pump is measured then by using pressure = force/area the pressure of the atmosphere can be estimated.

Apparatus required: •Bicycle pump •Slotted masses •Retort stand and clamp.

10 The hosepipe problem

Wrap a long piece of rubber tube round a can. Fit a funnel into one end and put the other end into a bucket. Hold the can horizontally in a clamp and pour some water into the funnel. You would expect water to come out of the other end and flow into the bucket but in fact the funnel overflows. No water ever emerges from the other end. The demonstration is more impressive if you can borrow a hosepipe wound round a reel. The funnel can then be a metre or two above the reel and still no water comes out of the other end. It works — or rather it doesn't work — because once you have some air trapped in a loop the build up of air pressure as more water is poured in prevents the water moving round the loop.

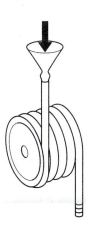

11 Lung pressure

This is a simple class experiment or demonstration of the pressure of your lungs. A large manometer filled with coloured water is all that is needed. Students blow into the lower end and see how far they can raise the water in the open end. Differences in levels of up to about two metres are usual with twelve-year-old students. To show the effect of exertion the experiment can be done before and after some form of physical exercise.

Apparatus required: Large manometer, up to 2 m high filled with
coloured water Use tubing connectors as mouthpieces, keeping them in
disinfectant to sterilize them before use.

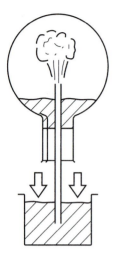

12 Atmospheric pressure: the fountain

Fit a glass tube through a bung fitted in the top of a round-bottomed flask
which is about a quarter full of water. To the end of the tube attach a short
length of rubber tubing which can be closed with a tube clip. Mount the
flask in a clamp so that it can be rotated. Heat the flask until there is steam
coming from the tube showing vigorous boiling. After boiling has been
going on for a few moments remove from the heat and immediately close
the clip. Then upend the flask so that the end of the tube is under the
water in the beaker and open the clip. As the pressure in the flask reduces
water spurts up the tube.

Apparatus required: •Round-bottomed flask with a glass tube fitted
through a rubber bung in its neck •Retort stand and clamp •Beaker of
water •Bunsen.

13 Bucket of chocolate blancmange (pudding)

Imagine a bucket full of chocolate blancmange (pudding) with a tight
fitting lid with a hole in the centre. The lid can be pushed down into the
bucket. As you push down on the lid the blancmange is forced up out of
the hole. This was originally supposed to be an analogy for a Fortin
barometer with the blancmange replacing the column of mercury.
However these days if the Fortin barometer is not a familiar instrument
the demonstration can simply be an analogy of how liquids are forced up
a tube when the pressure on the liquid outside the tube is greater than that
above the liquid in the tube.

14 Collapsing can and collapsing bottle

Both of these experiments are vivid demonstrations of the pressure of the
atmosphere.

(a) Put a little water in a tin can. Heat the can vigorously over a bunsen
until the water has been boiling for some time. Although there will be
some liquid left, the can will be full of steam. Remove the can from the
heat and quickly insert the stopper or screw on the lid. In the interests of
safety, don't heat a closed can!

As the can cools the steam inside the can condenses and therefore the
pressure drops. After a few moments the pressure difference between the
atmosphere outside and the small amount of residual air inside is enough

to crush the can flat! I usually stand on the top of the can first before starting the experiment to demonstrate how strong it is and then get a student to try and straighten out the can at the end after it has cooled.

A variation of this is to use an empty drinks can, put a little water in it and boil the water as before. Then pick it up with a pair of tongs and rapidly invert it in a bowl of water. The steam inside the can condenses, the small hole in the can prevents water from being sucked in too rapidly and the can collapses. ·

(b) This next demonstration of atmospheric pressure is very simple and direct and avoids heating cans of air! Completely fill a plastic squash bottle with water (bigger bottles are more impressive). Put a bung in it with a glass tube in the centre and attach a 2 m length of rubber tubing to the tube — more if the height of your lab will allow it. Get someone to hold the end of the tube closed while you climb on a bench and upend the bottle with the rubber tube dangling vertically downwards. Now open the lower end of the tube. As the water runs out the bottle will be squashed flat by the pressure of the air on the outside! The long tube gives a bigger pressure difference between the top and bottom of the water column and also prevents air leaking in.

Theory: pressure difference between the two ends of the water column of height $h = \rho g h$ where ρ is the density of the water.

Apparatus required: (a) •Tin can •Bunsen •Heat proof mat •Drinks can •Tongs (b) •Plastic bottle with rubber tube fitted to a bung in its neck •Water •Bucket.

---

15 Test tube in a bell jar

Take a large test tube filled with water and invert it in a beaker of water, making sure that the total volume of water present is less than the volume of the beaker. Place the arrangement in a bell jar connected to a vacuum pump and slowly reduce the pressure. The water level in the test tube will begin to fall as the pressure in the bell jar decreases, the water levels inside and outside the tube eventually becoming almost equal. (You need to be able to reduce the pressure below 0.01 atmospheres (10^3 Pa) to get much effect.)

If the air is allowed to leak slowly back into the bell jar the water level in the tube will rise again — a convincing demonstration that it is the air pressure on the water in the beaker that is forcing the water up the tube.

Apparatus required: •Vacuum pump •Bell jar •Large test tube •Beaker of water.

DENSITY, UPTHRUST AND ARCHIMEDES

General theory for this section

Archimedes principle states that: when an object is immersed in a fluid there is an upthrust which is equal to the weight of the fluid displaced. When the object is floating the upthrust = the weight of the object = the weight of fluid displaced = $V\rho g$ where V is the volume of the fluid displaced and ρ is the density of the fluid

If the density of the fluid is low then the object floats lower in the fluid so as to displace more of it.

1. Archimedes and solid rubber balls
2. Boats and Archimedes
3. Cartesian diver
4. Hippos
5. Ferries: displacement
6. Drinking straw hydrometer
7. Archimedes: Upthrust; top pan balance
8. Displacement and a bag of water
9. Upthrust

1   Archimedes and solid rubber balls

Find a rubber ball that just sinks in fresh water. Add salt to show that in the salty water it will begin to rise and eventually float. Experiment with the type of ball; a way of getting one that works is to drill a hole in a table tennis ball and add water until it just sinks; then close the hole with tape or wax.

Apparatus required: •Large plastic tank •Ball •Wax.

2   Boats and Archimedes

Use a toy boat with blocks of different materials in it floating in a tank to measure the change in water level if one or more of the blocks is removed or thrown into the water. My sons had a plastic boat about 15 cm long that was ideal for this; failing that, a simple foil cake tray may be used, although it doesn't look as good. You can use a TV camera to view what happens from the side. (See also the floating ice cube.)

Theory: there is no change of level if the blocks float when they are taken out of the boat but if they sink their density must be greater than water, they therefore displace a smaller volume of water than their own weight of water and so the level in the tank will fall.

Apparatus required: •Toy boat •Various blocks •Transparent plastic tank •TV camera if possible.

---

3 Cartesian diver

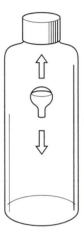

Blow a small bulb roughly 1 cm in diameter (smaller ones will work but they are not so easy to use or so impressive) on the end of a glass tube. After it has cooled cut it off leaving a short (0.5 cm) stem and then partly fill it with water. The actual amount needed will have to be done by trial and error but a little less than half full seems to work well. Then place it inside a large (2 litre) plastic drinks bottle which is filled almost full with water with the bulb at the top and put the top of the bottle on. Squeeze the bottle — the diver sinks because the increased pressure causes a decrease in the volume of air in the bulb, this reduces the upthrust by decreasing the volume of water displaced by the diver and the diver goes down. Try adding salt to the water to see the effect of an increase in the density.

Apparatus required: •Glass tube •Large plastic bottle •Bunsen •Heat resistant mat •Glass cutting knife •Water.

---

4 Hippos

I like to draw three hippos floating in water of different densities to show the change in immersed volume. They sink deeper as the density becomes less to retain the same upthrust. Mention the change in displacement depth as a ship goes from salty water to fresh water and the ease with which a person can float in the very salty waters of the Dead Sea. You may also notice the difference between a freshwater swimming pool and the sea — I can float in the sea without difficulty but find it almost impossible in the lower density fresh water in a swimming pool.

5 Ferries: displacement

It is helpful to use actual ferry data to teach a practical example of displacement. Ask the students to suggest the displacement depth of a 25 000 ton ferry and then calculate it in the following way.

Upthrust of a floating object = weight of object = weight of water displaced. The one I use as an example works out to have a displacement depth (draft) of about 6 m and it agrees with the P and O shipping line data (length 180 m, breadth 28 m, gross tonnage 28 000 tons).

6 Drinking straw hydrometer

Load a plastic drinking straw with lead shot and close the lower end with wax. It will now float upright in liquids. If it is calibrated in pure water it can then be used to measure the density of brine and methylated spirits.

Apparatus required: •Drinking straw •Modelling clay •Measuring cylinder or tall beaker and liquids.

7 Archimedes: upthrust; top pan balance

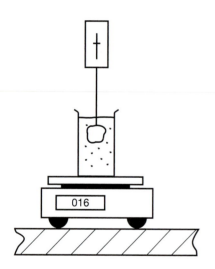

Put a beaker of water on a top pan balance and lower a lump of stone into the water. The stone should be suspended by a thread from a newton meter so that readings of the tension in the thread and the top pan balance can be compared as the stone is moved from being totally in air to totally in water. Just putting your finger into the water gives a good demonstration of increased reading on the top pan balance.

Theory: the reading on the spring balance decreases as the stone is lowered into the water due to the increasing upthrust as more of it is immersed. The reading on the top pan balance increases, the sum of both readings being constant and equal to the weight of the stone (if the weight of the container and water is allowed for).

Apparatus required: •Top pan balance •Beaker •Newton meter •Water •Stone •String or thread.

8 Displacement and a bag of water

Repeat experiment 7 using a bag of water instead of the stone. The readings on both the spring balance and the top pan balance should not change during the experiment since the weight of water displaced by the water is equal to the weight of the water itself.

9 Upthrust

A wonderful experiment can be carried out by dripping olive oil into alcohol. Put the olive oil into a funnel with a tap so that the outlet is below the surface of a large beaker of alcohol. Run some olive oil out slowly; the buoyancy of the alcohol 'cancels' the effect of gravity and allows the surface tension of the oil to pull it into circular bubbles. Really large bubbles can be made since the olive oil has very nearly the same density as the alcohol. Warming the alcohol will affect the buoyancy.

DANGER: do not set fire to the alcohol!

Apparatus required: •Funnel with tap •Large beaker •Alcohol •Olive oil.

MOTION

General theory for this section

Speed = distance/time (for steady speed).

Acceleration = change in speed/time taken.

Acceleration due to gravity at the Earth's surface (g) = 9.8 m/s^2 (often simplified to 10 m/s^2)

1. Animal Olympics sheets
2. Reaction time
3. Velocity — it's a vector; two people walking
4. Table tennis ball accelerometer
5. Resource suggestions
6. Horizontal and vertical motion

1 Animal Olympics sheets

I have a collection of sheets of speeds and athletic performances of various animals, birds and fish which I use at the beginning of work on speed and acceleration as a comparison with humans. They include a 100 m, 1500 m, marathon, long jump, high jump, swimming speeds and flying speeds.

Some data:

200 m	Man 25 mph; Ostrich 37 mph; Racehorse 44 mph; Cheetah 60 mph
1500 m	Man 16 mph; Gazelle 50 mph
Marathon	Man 12 mph; Racehorse 17 mph; Reindeer 25 mph; Antelope 36 mph
100 m freestyle	Man 5 mph; Salmon 20 mph; Wahoo 50 mph
Flying	Wasp 12 mph; Bee 20 mph; Partridge 55 mph; Ring necked duck 66 mph

Apparatus required: •Animal Olympics sheets.

2 Reaction time

This is a simple way of demonstrating and measuring the reaction time of a student. Hold a ruler vertically and ask one of the students to put their thumb and forefinger either side of the bottom of the ruler. Say 'Now' and drop the ruler. They have to catch it by closing their thumb and finger. Their reaction time (t) is simply measured by how far the ruler falls from rest (s).

An extension of this is to fix a piece of tape to the ruler and mark it with equal time intervals (0.1 s) using the formula below.

Theory: use $s = \frac{1}{2}gt^2$ to calculate t, where s is measured in metres and $g = 10 \text{ m s}^{-2}$. Reaction time $(t) = [2s/g]^{1/2}$

Apparatus required: •Ruler.

3 Velocity — it's a vector; two people walking

I use these ideas to introduce the vector nature of velocity. A vector is a quantity with both **size** and **direction**.

(a) Two students are placed at either side of the lab and asked to walk at, say, 1 m/s. In spite of having not been told to do so they invariably walk towards each other. I then get them to stand about two metres apart and ask them to repeat the procedure. It takes them a few moments to realize that the effects are quite different if one walks forward and the other walks backwards. I always mention Heathcliffe and Cathy in Wuthering Heights (the book would have been so different if they had walked the other way).

(b) As a further way of emphasizing the point consider the effect of a car waiting at traffic lights; if it goes into reverse instead of forwards when the lights change to green the effects can be unfortunate!

4 Table tennis ball accelerometer

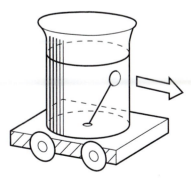

(a) Fix a table tennis ball to the lid of a glass jar by a thread. Fill the jar with water, put the lid on and then turn the jar upside down. The ball floats in the water. Now accelerate the jar, this can either be done on a linear air track or by placing the jar on a rotating table. The ball will swing in the direction that the jar is accelerating. In the case of the linear motion it will move in the direction of motion; in the second case it will move towards the centre, in the direction of the centripetal acceleration.

Alternative versions of this experiment can be carried out.

(b) Spirit level accelerated or spun on a turntable. The bubble goes to the forward part of the level or moves towards the end nearest the centre of the circle if it is rotated. Mount the spirit level on a truck on a linear air track and record the effects if possible using a TV camera for later analysis.

(c) Imagine a balloon filled with helium tethered by a string to the floor of a car. If the car accelerates the balloon of helium will move forwards since it has a lower inertia than the surrounding air.

Apparatus required: •Jar with screw top •Table tennis ball •Thread •Linear air track or rotating table •Motor.

5 Resource suggestions

Use data about cars and athletes to give a real everyday feel to the work of acceleration and motion. See also experiment 1.

100 m sprinter

Reaction time 0.109 s

Distance	10 m	30 m	70 m	100 m
Time	1.84 s	3.80 s	7.36 s	9.86 s
Average speed over previous 10 m	5.9 m s^{-1}	10.8 m s^{-1}	11.9 m s^{-1}	11.7 m s^{-1}

Family saloons	Mass (kg)	Time for 0–100 km hr^{-1} (s)	Power (kW)
A	906	12.7	51
B	875	11.8	55
C	1010	16.7	48
D	1750	7.8	150

6 Horizontal and vertical motion

(a) Get one of the class to sit on a wheeled chair. Push them across the lab and as they move ask them to throw a ball in the air. As far as they are concerned the ball goes straight up and comes straight down, but to the rest of the class it moves in a parabolic path. Making a large grid on the board behind the moving chair and taking a video of the motion makes this even clearer.

(b) An alternative to this experiment is to mount a metre rule on the top of a rider on a linear air track and drop a ball bearing from the top while the rider moves at a constant velocity. The ball bearing will always land on the rider.

Apparatus required: (a) •Wheeled chair •Ball •Large grid screen •TV camera if possible. (b) •Linear air track •Ball bearing •Ruler •Rider.

GRAVITY

General theory for this section

This section deals with experiments connected with the acceleration due to gravity (g), mostly that on the surface of our planet ($g = 9.8$ m s^{-2}). That means that if an object is dropped near the Earth's surface its speed increases by about 10 m s^{-1} every second if the effects of air resistance are ignored.

When an object falls from rest and accelerates under the effect of the Earth's gravity the distance it falls (h) in a time t is given by the equation: $h = \frac{1}{2}(gt^2)$. The gravitational field strength at the surface of the earth is approximately 9.81 N kg^{-1} and this will give a mass of 1 kg an acceleration of 9.81 m s^{-2}.

When a projectile is thrown it has a constant vertical acceleration (g) towards the ground but a constant horizontal velocity if we ignore air resistance. The horizontal and vertical motions are independent.

In some of the experiments a constant head apparatus is mentioned. This is simply a device for maintaining a constant head of water at an outlet.

1. Pearls in air
2. g with a water jet
3. Diluted gravity
4. Diluted gravity: projectile paths
5. g: gramophone turntable
6. The contracting stream
7. Falling can and water, what happens
8. Diluted gravity again
9. Two balls falling joined by stretched elastic
10. Falling helical spring
11. Falling bar method for g
12. Falling can with hole at one side
13. Monkey and hunter
14. Mooing carton
15. Vertical acceleration
16. Galileo inclined planes
17. Guinea and feather tube
18. Dropping books and paper: air resistance and drag

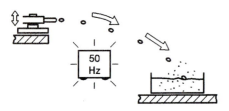

1 Pearls in air

This is a classic demonstration designed to show the parabolic path of projectiles in a gravitational field. A water jet is formed by using the glass part of a dropping pipette fixed to a thin walled rubber tube and connected to the water tap. The rubber tube is passed through an old style ticker timer or over a vibration generator so that the tube is alternately squeezed and released when the device is switched on.

The water jet falls in a parabola from an initial horizontal direction but is also interrupted by the pulsing so that droplets of water are formed instead of a continuous stream. If the arrangement is illuminated with a stroboscope, pearl-like droplets of water can be made to stand still or move slowly through the air. The constant horizontal velocity and the increasing vertical velocity can be seen by observing the positions of successive drops. To get a permanent record you could mark the position of the shadows of the water drops on a screen behind the jet or even photograph it. A truly beautiful demonstration.

An extension of the basic version is what I call the **double pearls in air**. In this experiment two jets are used from different water taps but with tubes running under the same ticker timer bar. One is adjusted to give a parabola while water simply trickles out from the other, falling vertically. The vertical acceleration of the drops can then be compared. Of course you can make two parabolas and compare these.

Warnings about pupils and flashing lights should be given here.

Theory: since $h = \frac{1}{2}gt^2$ and $s = vt$ the equation for the parabolic path for the water is $h = gs^2/2v^2$ where s is the horizontal distance travelled, h the vertical distance and v the horizontal velocity of the jet

Apparatus required: •Ticker timer •Two water jets •Constant head apparatus •Bucket •Stroboscope.

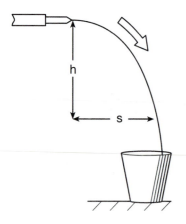

2 g with a water jet

The value of the acceleration due to gravity (g) can be found in a rather novel way by using a jet of water projected horizontally from a dropper attached to a constant head to give a parabolic path. The shape of the path is found by measuring pairs of values of the height fallen (h) and the distance (s) horizontally from the orifice and if the rate of flow of the water is also found the value of g can be calculated. Measure the diameter of the jet to calculate its cross-sectional area (A).

The horizontal velocity (v) is obtained from the equation $V = Av$ where V is the volume of water leaving the dropper per second (measure this by directing the jet into a measuring cylinder) and A is the cross sectional area of the jet. Using a TV camera to give an image on the screen or shining light from a projector to make a shadow of the path on a board are both helpful ways of making measurements easier to take.

Theory: $s = vt$; $h = \frac{1}{2}gt^2$; $v = V/\pi r^2$

Apparatus required: •Water jet •Constant head apparatus •Rulers •Base clamp •Measuring cylinder •Stop clock •Travelling microscope or vernier or TV camera •Bucket •Mop!

---

3 Diluted gravity

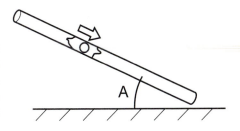

(a) Realizing the problem of making accurate measurements of the acceleration due to gravity, Galileo diluted gravity by rolling balls down slopes. His original apparatus is in the History of Science Museum, Florence. We can recreate his experiment by rolling a marble down an inclined plastic ramp or tube and measuring the time it takes to travel a measured distance.

The gravitational acceleration (g) has been 'diluted' to $g \sin A$ where A is the angle that the tube makes with the horizontal. It is important to understand that it is much more accurate to measure the small angles by trigonometry than by fiddling around with a protractor!

Carrying out the experiment by using a rider on a tilted linear air track (I use one 2 m long) can give extremely accurate values for g.

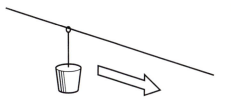

(b) An alternative version of the diluted gravity experiment of Galileo can be performed on a large scale with an aerial ropeway-type arrangement fixed across the lab. A wire should be fixed tightly from a high point on one side of the lab to a low point on the other. A small cup either fixed to a pulley wheel or simply tied to a loop of wire can then travel down the wire. Time, distance and angle can easily be measured.

Theory: acceleration down the plank or wire = $g \sin A$; $s = \frac{1}{2} g \sin A \, t^2$

Apparatus required: (a) •Wooden ramp and track or plastic tube •Marble •Stop clock •Ruler (b) •Wire •Cup and pulley wheel.

---

4 Diluted gravity: projectile paths

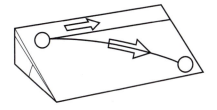

An extension of the diluted gravity experiment (see experiment 3) is to investigate a diluted projectile path. Fix a large sheet of paper to a drawing board. On top of this fix a piece of carbon paper, face downwards. Tilt the board and then roll a heavy ball bearing across the top of the paper in a horizontal direction. The path of the ball bearing will be produced on the paper. Different angles of tilt and different path directions can be used. This would be suitable for an introduction to projectiles or at a more advanced level where calculation of the parameters of the paths can be performed.

Apparatus required: •Drawing board •Large ball bearing •Carbon paper •White paper.

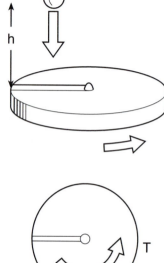

5 *g*: gramophone turntable

A rather quaint experiment is the use of an old gramophone turntable to measure the acceleration due to gravity (*g*). The problem with all such measurements is to find a way of determining the time of fall which will always be rather small over the distances possible in a laboratory. In this method, this small time is found by using a gramophone turntable. First fix a piece of tape along a radius. Hold a ball bearing at height *h* above the rotating turntable and release it at just the moment when the tape passes beneath it. The angle through which the turntable has rotated before the ball bearing hits it is found by either covering the surface with modelling clay or a piece of carbon paper over a white sheet of paper. The period of rotation of the turntable is determined using a stopwatch and may be used to calculate the time of fall (*t*). The acceleration due to gravity is then worked out from $g = 2h/t^2$. Admittedly, it is very inaccurate but it does give a means of obtaining *g* and then commenting on why it would be an unreliable answer.

Apparatus required: •Gramophone turntable •Large ball bearing •Carbon paper and white paper or modelling clay •Ruler.

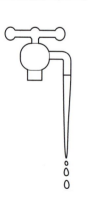

6 The contracting stream

The speed of a jet of water falling vertically from a tap into a sink increases the further from the tap it gets. This would seem to suggest that more water reaches the sink every second than is being emitted from the tap. Clearly impossible! This can only be explained if the stream of water gets thinner with increasing depth below the tap. This can be verified by turning the tap on slightly and observing the stream.

7 Falling can and water: what happens

Drill a hole in the bottom of a tin can. The size isn't critical but two or three millimetres in diameter will be fine. Put your finger over the hole and fill the can with water. Now drop the can; the water stays inside. This is much as you would expect, since all objects accelerate downwards at the same rate if air resistance is ignored. Now repeat the experiment but drop the can after you have allowed some of the water to start streaming out. What happens to the water? It looks as if the can is continuing to empty itself, but this would mean that the water is falling with a greater acceleration than g. This is impossible of course! The can and the water both accelerate at the same rate, g, and the can has the same amount of water in it when it reaches the ground as it had at the start of the drop.

Apparatus required: •Bowl or bucket •Tin can with hole.

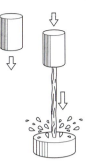

8 Diluted gravity again

Another variation of the diluted gravity experiment (see experiment 3) is to use a 30 cm long clear plastic ruler that has a groove down its centre, a ball bearing and an overhead projector. Put the ruler on the overhead projector with one end slightly raised (a millimetre or two). (You may need to support the ruler in the middle to stop it bowing.) Now let the ball bearing roll down it; using the image of the ruler on a screen to show the distance reached at certain time intervals. Calculate the acceleration due to gravity as in experiment 3.

Apparatus required: •Overhead projector •Ball bearing •30 cm clear plastic ruler •Stop clock.

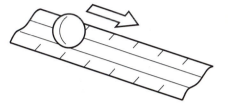

9 Two balls falling joined by stretched elastic

An interesting problem involving gravity is to take two balls that are joined together by a piece of stretched elastic, and hold one of them so that the other hangs below it, the elastic between them being stretched. Now release them so that they fall. What happens to their separation as they fall? It is worth repeating the experiment both with balls of the same mass and of different masses and trying it with the greater mass at either the top or bottom.

Theory: The upper ball falls with a greater acceleration than the other; the two are pulled together by the elastic and so the acceleration varies until the elastic becomes slack when they both fall with an acceleration of g.

Apparatus required: •Two power balls •Piece of elastic.

10 Falling helical spring

A variation of experiment 9 is to drop an extended helical spring and observe what happens to various parts of it as it falls. You will find that during the drop the bottom coils stay where they are while the upper coils catch up with them and then the whole spring falls together. During the whole motion the centre of mass falls with an acceleration of g.

Using a TV camera to record the fall and looking at a slow motion replay will make the results of both experiments 9 and 10 much easier to appreciate.

Apparatus required: •Helical spring •TV camera if possible.

11 Falling bar method for *g*

You can use the fact that the vertical acceleration of any point on any rigid falling object is the same, no matter whether it is dropped vertically or swung or projected at an angle, in the following experiment to find *g*. A metre ruler is pivoted at one end and held at an angle by a thread fixed to its lower end, the thread being looped over the pivot bar and with a sufficiently heavy pendulum bob tied to the other end. Now burn through or cut the thread. The ball begins to fall and the ruler begins to swing downwards at the same moment. The position where the ball meets the bar can be used to find *g*. Finding this position can be made easier by putting a piece of carbon paper over a strip of white paper that is fixed to the ruler.

Theory: since the ball bearing hits the ruler vertically below the pivot the time taken for the fall will be one quarter of the period of oscillation of the ruler. The period can be found by measuring the time for ten swings of the ruler and then working out the time for one quarter of a swing.

Apparatus required: •Pivoted metre ruler •Retort stand and clamp •Ball bearing •Thread •Matches •Stop clock •Carbon paper •White paper.

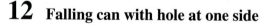

12 Falling can with hole at one side

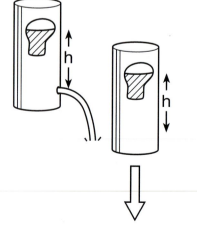

The can with holes (see liquid pressure, experiment 1) can be used to demonstrate that if there is no gravitational attraction there will be no liquid pressure. For this experiment use a can with just one hole in one side near the bottom. Fill it with water, cover the hole with your finger and then drop it. Since both can and water fall together there is no net gravitational force and so the water stays in the can.

Theory: pressure in a point in a liquid = $h\rho g$ and since the net value of *g* is zero for the falling can and water there is no pressure difference between the top and bottom of the water in the can.

Apparatus required: •Tin can with hole near the bottom •Water.

13 Monkey and hunter

A monkey hangs from a tree in a jungle and is discovered by a hunter who decides to shoot it. Pointing the rifle between the eyes of the monkey he prepares to pull the trigger. The monkey, being fairly intelligent, reasons that if he waits until the moment the bullet leaves the barrel and then drops out of the tree the bullet will pass over his head. The hunter pulls the trigger, the monkey waits until the bullet is leaving the barrel and lets go. To his dismay the bullet hits him directly between the eyes! He was intelligent but had forgotten his physics!

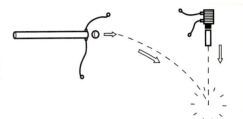

The explanation for this can be demonstrated by a classic experiment that shows the constancy of acceleration for falling bodies. Mount an electromagnet in a clamp about 0.5 m above a bench and mount a blowpipe horizontally in another clamp so that it is pointing just below the core of the electromagnet. Put a marble in the blowpipe, fix a small strip of aluminium foil across the mouth of the blowpipe and then connect up a series circuit with the electromagnet, a dc power source and the aluminium strip. Switch on and hang a tin from the electromagnet making sure that the blowpipe is pointing at the centre of the tin. Blow sharply down the pipe and the marble will fly out, breaking the foil, causing the tin to fall. The marble will fall at the same rate as the tin and should collide with it before hitting the bench. I have once or twice managed to hit a falling ball bearing. (see water path in a gravitational field, experiments 1 and 2.)

Apparatus required: •Blow pipe •Marble •Electromagnet •Tin lid
•Aluminium foil •Power supply.

14 Mooing carton

The mooing milk carton can be used as a fun problem to show the constancy of vertical acceleration in free fall and also to demonstrate g forces. Turn it upside down and drop it in 'mid moo'. Observe the change in the sound as it goes down. The mooing stops in free fall and starts again when the high deceleration forces occur as it is caught.

Apparatus required: •Mooing milk carton.

15 Vertical acceleration

The 'feel' of the value of the acceleration due to gravity can be gained by putting a small object such as a ball bearing on your hand and then moving your hand downwards. If you move it with an acceleration of less than g the ball bearing stays in contact with your hand but if your hand accelerates with a greater acceleration than g the ball bearing leaves the surface. It is rather more difficult to do this with your hand on top of the object. You can compare this with looping the loop in a rollercoaster or with people in a car going over a bumpy road. You will leave your seat in a car if it travels over the bumps too rapidly.

16   Galileo inclined planes

An interesting effect of the acceleration along inclined planes can be shown by a variation of Galileo's experiment on diluted gravity. Thread a bead onto each of a set of wires starting at one point on a vertical bicycle wheel from which the spokes have been removed and ending at different points along the circumference. When the beads are released from the top they slide down the wires keeping a circular arrangement and all reaching the end of the chord at the same time. A related problem in gravitation refers to the fact that it takes 42 minutes for objects falling through holes in the Earth to reach the other side whatever chord is used (this is of course a theoretical and ideal situation and ignores all frictional effects). It would make an ideal and rapid transport system. You can extend the idea to SHM where the body is free to oscillate about the centre of the Earth. Students often find it difficult to accept that the acceleration is zero at the centre of the 'fall'.

Apparatus required: •Bicycle wheel with spokes removed and wires fitted with beads on them.

The following two experiments show the effect of air resistance on falling objects.

17   Guinea and feather tube

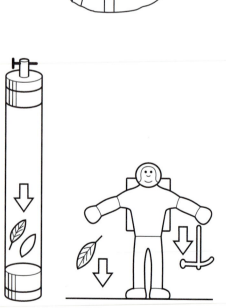

This is a classic experiment to show the effect of air resistance and the constancy of the acceleration due to gravity. Take a 1 m long glass tube of diameter about 5 cm, put a small piece of feather and a penny into the tube and fit bungs tightly into both ends, one with a metal tube in the centre. Attach the tube to a vacuum pump. Upend the tube and show that the penny falls faster than the feather because it has much lower air resistance. Now pump out the air and show that they both fall at the same rate. A video clip of astronauts dropping a falcon feather and a hammer on the Moon illustrates this as well. (It is important to realize that on the Moon there is no air but there is still a gravitational field, about 1/6 of that at the surface of the Earth.) It is certainly not true to say that no air means no gravity.

Apparatus required: •Guinea and feather tube •Vacuum pump •Coin •Feather.

Safety consideration. Put some sticky tape around the lower few centimetres of the tube to prevent the tube shattering if the penny hits it too hard!

18 Dropping books and paper: air resistance and drag

This is an interesting experiment on air friction but it is important to stop between each part and ask the students what happens next.

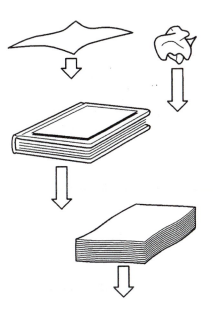

(a) Drop a sheet of paper; it falls slowly due to air friction.
(b) Now crumple it up; its mass is unaltered but the crumpling reduces the air friction and it falls quicker.
(c) Then use another similar flat sheet of paper but this time with a book on top of it — the effect of the air friction on the paper is removed.
(d) Then drop a sheet of paper with a book underneath it; they both fall together.
(e) Finally, drop a ream of loose paper. All the sheets fall at the same rate.

An alternative to parts (c) and (d) is to use a metal disc with a similar sized paper disc placed either on top of it or below it.

These experiments remove the need for the traditional guinea and feather experiment if you don't have a vacuum pump.

Apparatus required: •A stack of loose paper •A book of similar area.

A further experiment about gravity involves a falling candle and is described in the section on convection (experiment 6).

NEWTON'S LAWS — MASS AND ACCELERATION

General theory for this section

Newton's three laws of motion may be summarized as follows.

1. A body remains at rest or in a state of uniform motion unless acted on by a force. What is still stays still and what is moving stays moving at a steady speed in a straight line unless a force acts on it.

2. The resultant force on a body causes an acceleration which depends on both the size of the force and the mass of the body, i.e. force = mass × acceleration

3. If a force acts on one body an equal and opposite force acts on another body.

1. Pushing and Newton's Third Law
2. Sling shot theory
3. Newton's Laws
4. Knees bend on bathroom scales
5. Linear air track and two pulleys
6. Shopping bag and old lady: acceleration
7. The lift problem
8. Newton's Third Law

1 Pushing and Newton's Third Law

This is a very simple example of Newton's Third Law. Get two students of similar size to stand facing each other. Tell them to raise their hands so that the pairs of hands are touching. Then tell one student to push the other over. They will both be pushed backwards. This example of Newton's Third Law confirms that when a force acts on one body an equal and opposite force acts on another body. Trying the same thing with both students on skateboards may be even more impressive.

Apparatus required: •Two skateboards (optional.)

2 Sling shot theory

When a space craft approaches a planet that is moving through space the spacecraft experiences a sling shot effect due to the motion of the planet whose velocity is only altered to an infinitesimal extent. As the spacecraft approaches the planet the gravitational attraction between them will increase the velocity of the spacecraft so that when it passes the planet it is moving with a greater velocity relative to the planet. However, since the planet itself is also moving through space, the absolute velocity of the spacecraft is also increased.

3 Newton's Laws

A simple experiment for the verification of Newton's Second Law uses a small trolley which is accelerated along a friction compensated track by a number of washers tied to a thread which is fixed to the trolley after passing over a pulley. A way of ensuring the constancy of mass when studying Newton's Laws is to transfer the washers from the trolley to the accelerating mass or vice versa. This is a useful tip when plotting graphs of acceleration against force (the total mass of all objects accelerating is thus kept constant).

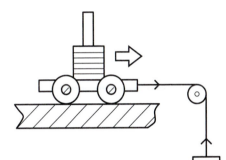

Theory: Accelerating force $(F) = mg$ = total mass $\times$ acceleration = $(M + m)a$ where m is the mass of the washers accelerating the trolley and M is the total mass of the trolley and its load.

Apparatus required: •Trolley •Runway •Bench •Pulley •Washers •Thread •Ruler and stop clock or light gate assembly and timer.

4 Knees bend on bathroom scales

You can observe some very interesting effects by standing on a set of bathroom scales and then watching the reading as you bend your knees. The reading goes down as you bend your knees but increases as you straighten your legs to accelerate yourself upwards to a standing position. An alternative version is to stand on the scales holding a pair of dumbbells in your hands and watch the reading as you raise and lower them.

Apparatus required: •Bathroom scales •Two dumbbells or weights on a bar.

5 Linear air track and two pulleys

The effect of two unbalanced forces is very well demonstrated by having a trolley attached to two masses each hanging over the opposite ends of a linear air track. The resulting acceleration due to the resultant force can therefore be found. (To avoid the air hose pass one thread over a pulley on an adjacent bench.)

A simplified version is just to use two dissimilar masses (M and m) hanging by a thread over two pulleys fixed to retort stands with a central mass held in mid-air. If this mass is not too great the string will be more or less horizontal and the applied forces will then be horizontal also.

Theory: the resultant accelerating force (F) is given by the equation $F = Mg - mg = (M + m)a$, where a is the resulting acceleration.

Apparatus required: •Linear air track, blower and trolley •Two pulleys •Two retort stands •Two sets of slotted masses.

6 Shopping bag and old lady — acceleration

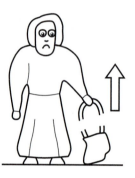

Why is it that you can lift a shopping bag full of shopping slowly without a problem but if you lift it quickly the handles break? Think about an old lady picking up her shopping bag quickly. The handles break because they have to produce an acceleration as well as simply support the weight of the bag. This can easily be demonstrated by putting a heavy weight inside a plastic shopping bag and showing that it will just support the load but that it breaks — usually at the bottom — when accelerated briskly upwards.

Another example of this is the maximum acceleration allowed by a pilot of a rescue helicopter when lifting off a stretcher from a hillside before the supporting ropes break.

Theory: for the shopping bag: force in handles = weight of shopping + force to accelerate the bag upwards ($F = mg + ma$).

For the rescue helicopter: If the injured climber and the stretcher had a combined mass of 180 kg and the breaking force of the rope supporting them was 2000 N then things would be fine if the helicopter just hovered ($mg = 180 \times 9.8 = 1764$ N). However if the helicopter pilot tried to accelerate upwards at a mere 2 m s^{-2} the result would be disastrous ($mg + ma = 1764 + 180 \times 2 = 2124$ N, breaking the rope).

Apparatus required: •Two 5 kg masses •Plastic shopping bag.

7 The lift problem

Make a table to discuss what happens to your mass, weight and reaction when you are in a lift that stays still, moves with constant velocity up or down, accelerates up or down or falls freely. Remember that as long as the gravitational field does not change your weight does not alter — it is only the reaction of the floor that may change if you accelerate up or down.

8 Newton's Third Law

Mount a small bow on a dynamics trolley and fire a rubber-tipped arrow from it. As the arrow goes forward the bow and frame accelerate backwards. This is an interesting and useful model with which to demonstrate both Newton's Third Law and the transfer of energy from that stored in the stretched bowstring to the kinetic energy of the trolley and the arrow.

Apparatus required: •Dynamics trolley •Toy bow •Rubber-tipped arrow.

WORK, ENERGY AND POWER

General theory for this section

Work done = Energy changed from one form to another = force ×
vector displacement

Power = energy transformed/time taken

1. Energy loss in a rubber band
2. Cars and carpet
3. Power of a student — running upstairs
4. Train sheets

1 Energy loss in a rubber band

Take a rubber band and cut it so that it can be used as a single strip.
Stretch the band by adding weights and then remove them carefully. The
resulting energy loss can clearly be seen from a graph of load against
extension (a useful assessed practical). The area between the loading and
unloading lines is the energy lost within the band due to the heat
generated by the twisting and untwisting of the long molecules. We use
bands about 3 mm wide and load them to around 15 N giving a maximum
extension of 50 cm. (See also the section on elasticity.)

Apparatus required: •Rubber band •Retort stand and clamp •Slotted
masses (up to 15 × 100 g)

2 Cars and carpet

This experiment may be used to measure the braking force of a toy car by
allowing it to run down a hardboard ramp onto a piece of carpet. Measure
the stopping distance and hence find the braking force. This can be done
by either measuring the loss of potential energy as it runs down the ramp
and onto the carpet, or the loss of kinetic energy by measuring its speed
with a light gate as it reaches the carpet. (This is good for an
investigation-type experiment.)

Students may like to suggest which is the more accurate version.
Braking forces of around 0.05 N are obtained with the cars I have used.
(The method using the loss of potential energy is the easier of the two for
younger students). Compare this experiment with the use of escape lanes

filled with gravel at the sides of steep hills. A graph of velocity squared against braking distance shows the dependence of kinetic energy on v^2 and not simply v.

Theory for this experiment: For a car of mass m travelling down a ramp from a height h and reaching a speed v at the bottom, loss of kinetic energy as it brakes to a stop on the carpet = braking force (F) × braking distance (s).

Potential energy lost in travelling down the ramp = mgh. If we assume that this is all converted to kinetic energy then $mgh = Fs$.

However, if we use the second method we can allow for energy lost on the ramp.

Kinetic energy (measured directly with a light gate) = $\frac{1}{2}mv^2 = Fs$

Apparatus required: •Toy car •Piece of carpet (not too rough and about 0.5 m long) •Hardboard ramp (about 0.4 m long) •Ruler.

3 Power of a student running upstairs

This is a simple experiment to measure a student's power; they carry their own mass (themselves) up a known height. The students run up a flight of stairs of known height and measure the time that they take to do it. They then measure their own weight and so calculate the work done and the power that they developed. They can also try it carrying one of the other members of the class if you think it is safe. Alternatively, carrying a rucksack loaded with books will increase their weight.

The power of their arms can be found by lifting weights or using a multi-gym. Don't forget that in some versions of this they have to lift their arms as well; how do you find the mass of an arm? Warn them about the dangers of dropping a heavy weight on their head!

Apparatus required: •Bathroom scales •Flight of stairs •Stop watch •Measuring tape or ruler •Heavy mass (5 kg).

4 Train sheets

This is a set of sheets devised to teach the topics of work, force and energy. It introduces the idea of the energy stored in a spring, the resistive forces opposing motion and the resulting stopping distances. We have three trains with one, two and three units of stored energy and these are run on three different surfaces having one, two and three units of frictional resistance. If the train with one unit of stored energy travels 30 m on a surface with one unit of resistive force before stopping then the students are asked to work out how far all the trains would travel on all the surfaces.

The end result is hopefully an understanding of the formula work done = energy transferred = resistive force (F) × distance travelled (d).

Having a clockwork train for an actual demonstration helps!

Apparatus required: •Train sheets sample

Energy	Force (F)	Distance travelled (d)	Force × Distance
1	1	30	30
2	1	60	60
3	1	90	90
1	2	15	30
1	3	10	30

ROCKETS

1. Carbon dioxide rocket
2. Water powered rockets
3. The firework rocket — upside down!
4. Balloon rocket

1 Carbon dioxide rocket

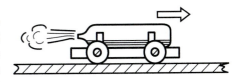

The carbon dioxide rocket trolley is a superb piece of apparatus for the study of momentum. It is simply a sparklet soda siphon bulb mounted horizontally on a small trolley. The side of the trolley is fixed to a piece of string that is looped round a retort stand on the floor. The idea is to use a pair of compasses with a short point to puncture the sparklet bulb allowing the carbon dioxide gas to rush out — the reaction on the bulb making the trolley move off in the opposite direction — 'orbiting' the retort stand at high speed!

The size of the hole is critical: too big and the gas comes out rapidly and it is all over too soon, too small and the result is a rather pathetic movement. Measuring the rotation rate in a circle of known radius can give the average velocity of the trolley.

Putting the trolley on a set of rails along the lab has produced some spectacular accelerations, but safety must be considered in case it leaves the track!

Apparatus required: •Carbon dioxide rocket trolley •Spartlet soda siphon bulbs •Pair of compasses or dividers •String •Retort stand •Rulers •Stop clock.

2 Water-powered rockets

There are two types and both provide excellent demonstrations.

(a) A small plastic rocket that is filled with a little water, mounted on its pump mechanism and the air is pumped in. When sufficient air has been put in the rocket is launched; the high air pressure within the rocket ejects a stream of water and the rocket flies up.

(b) The second, and rather better, version is simply a plastic drinks bottle with a set of fins and a valve; again air is pumped in but this time the valve automatically releases the rocket when the pressure inside the bottle reaches a certain amount.

Apparatus required: (a) •Water rocket and pump mechanism (b) •Plastic drinks bottle •Valve and fins assembly •Bicycle pump.

3 The firework rocket — upside down!

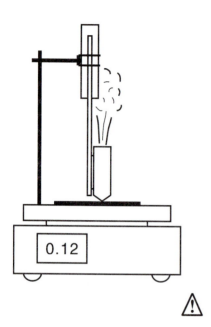

This needs to be done outside the lab and with great care. Mount a small firework rocket, one without star shells, pointing downwards and held loosely in a tube in a clamp on the top of a cheap top pan balance reading to an accuracy of one gram. Light the rocket and stand well back behind a safety screen (this is essential). The thrust of the rocket acts downwards and the reading on the top pan balance gives the thrust of the rocket during the firing. Use a TV camera if possible to record this for later analysis. The 'toy' chemical rockets that are produced are sold with data sheets that give good force against time curves and these can be used for analysis if you do not wish to do the rocket experiment yourselves!

Some safer alternatives would be to use a balloon, a soda siphon bulb or a plastic bottle containing dilute acid and chalk.

Apparatus required: •Firework rocket (small, without star shells) •Top pan balance •Small metal tin lid •Piece of hardboard to protect balance •Safety screen •Retort stand and two clamps •Launch tube to hold rocket.

4 Balloon rocket

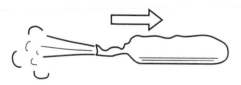

(a) Blow up a balloon and then release it so that it flies around the lab. This is a classic demonstration of Newton's Third Law and the conservation of momentum. The momentum of the exhaust air is equal and opposite to that of the rubber of the balloon. The momentum of the complete system before release is zero and that after release must therefore also be zero. I have found that it works better with a sausage-shaped balloon.

(b) A variation of this simple experiment is to use a 'tethered' rocket by fixing the balloon to a plastic straw which is threaded on a taut line across the lab. Keep the balloon closed with a clothes peg while you prepare the mount.

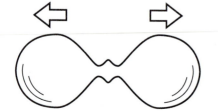

(c) The principle of action and reaction can be demonstrated by a variation of the balloon rocket. You will need two balloons: one with a short length of plastic tube (a centimetre or so) fitted in its neck. Now blow up the other balloon, close off its neck with a clip and fix it to the other end of the tube. Now open the clip. The inflated balloon does not fly around the lab — the emitted air produces a force on the other balloon which prevents it from moving.

(d) Another demonstration of this principle is to take a plastic straw, one with a flexible section and bend it into an L shape. Pivot the top and blow into it through a tube; the emitted air will make the straw turn. Now fit a small plastic bag or a piece of cling film over the end. No air escapes; there is a force on the bag and the straw does not turn.

Apparatus required: •Balloons — at least one sausage shaped one •Plastic straw •Tight wire mounting •Clothes peg •Plastic tube.

MOMENTUM, COLLISIONS AND EXPLOSIONS

General theory for this section

This section of experiments deals with the applications and effects of momentum.

Momentum = mass $\times$ velocity

Momentum is conserved in all collisions whether they are elastic or not, e.g. $m_1u_1 + m_2u_2 = m_1v_1 + m_2v_2$

If an impulse (force $\times$ time) is given to an object the object undergoes a change of momentum, e.g. $Ft = mv - mu$

1. Elastic collisions
2. Newton's cradle
3. Snooker and Newton's cradle
4. Momentum in elastic and inelastic collisions
5. Momentum and snooker/football game
6. Two power balls — dropped on top of each other
7. Smacking hands
8. Kicking a football
9. The air rifle and momentum
10. Toy gun on linear air track
11. Stopping an escalator
12. Pendulum on a trolley: momentum conservation
13. Sand falling onto a top pan balance
14. Crumpling zones in cars
15. Collisions
16. Momentum in catching
17. Helicopter details
18. Throwing and jumping: momentum/inertia
19. Force in a long jump take off or when a ball bounces
20. Momentum and the rotating table

1 Elastic collisions

In an elastic collision no kinetic energy is lost and this is very difficult to show in practice since elimination of friction is a problem. The ideal case can be approached by using two magnets. One magnet should be mounted horizontally in a block or held in a clamp. The second magnet is then suspended on four threads so that it is restricted to swing in one vertical plane and hangs with one pole facing the similar pole of the fixed magnet. It is then pulled out and released. It swings in towards the fixed magnet, is repelled and so swings back and forward and keeps going for a long time. It is worth referring to the collisions between gas molecules in kinetic theory as being virtually perfect elastic collisions.

Apparatus required: •Two strong bar magnets •Thread •Wooden retort stand with two clamps and bosses.

2 Newton's cradle

A beautiful example of collisions: try pulling out a different number of balls each time and letting them swing back inwards. You will find that however many balls you pull back, up to four out of five, the same number will always swing out at the opposite end. Pulling one ball out from each end will give a ball bouncing in and out at either end.

Theory: at collision the impact is very nearly elastic and the momentum and energy of the swinging ball are transferred to the adjacent ball. This process is repeated along the line until the free ball at the end swings away.

Apparatus required: •Newton's cradle •TV camera and video if possible.

3 Snooker and Newton's cradle

A simple way of explaining Newton's cradle is to use snooker balls. Extend from just the two ball example where one ball stops and the other moves off (see experiment 5) to more and more balls. Each time it is only the end one that moves off (allowing for spin and friction, of course).

Apparatus required: •Snooker balls •Snooker table if possible.

4 Momentum in elastic and inelastic collisions

Set up a block on the bench and hang a power ball from a thread so that it just touches the block. Pull it up and allow it to swing back. The rebound involves a direction change and hence a large change of momentum. This means a large force and so the block topples. Repeat with a ball of soft modelling clay loaded to give the same mass. The block stays upright. This time the change of momentum is only half that of the rebounding ball.

Theory: change of momentum = Ft and so if the momentum change is large (for a given stopping time) the force is also large.

Apparatus required: •Power ball •Wooden block •Modelling clay •Thread •Retort stand.

5　Momentum and snooker/football game

Use a snooker or billiards table to demonstrate momentum conservation. (You don't need a real snooker table for this, just the balls and some simple form of cue on the lab bench.) In some ways the sliding may improve the result. Show that if a stationary ball is hit by a moving ball the first one moves off while the second one comes to rest. Also measure the angle between direction of travel of the two balls after an oblique collision — it really does come out to be about 90° with balls of equal mass. This should be related to the collision between particles of equal mass in nuclear interactions. Mention the problem of spin. It works reasonably well if you undercut the strokes.

　　Trick shots: it is possible to pot the last of a line of balls as long as the last two are set up in line with the pocket.

Apparatus required: •Snooker balls and snooker table if possible.

---

6. Two power balls — dropped on top of each other

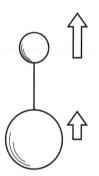

This can be bought as a demonstration experiment but simply using two power balls, one large and one small will do. Hold the balls above the ground with the small one resting on top of the large one. Now drop them so that they fall and hit the ground. The small one on top will leap off. It's even better with more balls. A needle in the lower ball with a thread through a hole in the top ball keeps the whole thing vertical.

　　I recall being told that twelve balls stacked on top of each other would theoretically put the top one into orbit; indeed, it would escape from the solar system! This is not only a good example of momentum conservation but also of the distribution of energy in explosions.

Theory: in a collision or explosion the momentum before collision or explosion = the momentum after collision or explosion.

　　Imagine that both balls are falling with velocity u at the instant the large ball hits the ground. So as the large ball rebounds from the ground at velocity V the small ball hits it at relative velocity $u+V$. If the collision is perfectly elastic it rebounds with relative velocity $u+V$, but since the large ball is moving upward with a velocity V the actual vertical velocity of the small ball is $u+2V$.

Apparatus required: •Two power balls, one large and one small •Thread •Needle.

7 Smacking hands

This is a simple but good example of Newton's Third Law: when a force acts on a body an equal and opposite force acts on another body. I suggest that you ask a pair of students to do this. I received a nasty blow by asking one of the girls to smack my hand once — it really stung. Get one of them to put their hand out and hold it still while the other one strikes it. You can extend the idea further by asking the students what it would be like to punch the wall. The force on your knuckles would be the same as that on the wall but the resulting effects and damage would be quite different!

Apparatus required: •Two students and their hands!

8 Kicking a football

This classic experiment is used to consider the momentum change of the football and hence the force used to kick it. A piece of aluminium foil is taped to the football and another piece is fixed to the shoe of the kicker. The foil on the foot is connected by a long lead to one of the start terminals of a scaler, while the other terminal is connected by another long lead to the foil on the ball. The ball is placed on the edge of the bench a known height above the ground and kicked off horizontally, the scaler recording the time for which the foot was in contact with the ball and the connecting wires breaking as the ball moves off. The horizontal distance travelled is measured together with the time taken to hit the ground and from this the horizontal speed of the ball is found. From the values of mass (m), velocity (v) and time (t) the force used to kick the ball can be found. I have tried it outside with a hockey ball. This works fine if the wires are allowed to separate after the ball has moved a short distance and somebody hangs onto the scaler!

Theory: horizontally $s = uT$ and $Ft = mu$. Vertically $h = \frac{1}{2}(gT^2)$ where t is the time of contact between the foot and the ball and T is the time taken for the ball to reach the ground having fallen a height h and travelled horizontal distance s.

Apparatus required: •Football •Scaler •Crocodile clips •Long leads •Aluminium foil •Measuring tape (10 m) •Tape.

9 The air rifle and momentum

How safe are these two experiments in a school laboratory today? I believe that if done with care and due regard to everyone's safety they are fine. I would like to include them.

(a) Fix two timing gates — simply a plastic frame with a strip of aluminium foil across the centre — a metre apart and connected to a scaler. Fire an air rifle pellet through the two gates so that it starts the timer when the first strip of aluminium foil is broken and stops it when the second is broken. The time for the pellet to travel the one metre between the two pieces of foil is given directly and the speed *v* of the bullet can easily be worked out

A means of catching the pellet should be found; a box of polystyrene backed with a wooden board is suitable for this.

(b) Using the same mounted air rifle, fire a pellet into a block of plasticine mounted on a model railway truck on rails. The runway should be tilted to compensate for friction and the speed of the truck (*V*) after impact should be found by timing its first 20 cm of movement.

 This experiment must not be performed unless the air rifle is securely bolted to the baseboard of the track.

Theory: knowing the mass of a pellet (*m*) and the mass of the truck and modelling clay (*M*) the speed of the pellet can be found ($mv + MV = 0$)

Apparatus required: (a) •Timing gates as described (b) •Mounted air rifle •Truck loaded with modelling clay •Scaler •Ruler •Stopwatch.

---

10   Toy gun on linear air track

The recoil of a rifle can be demonstrated by using a toy gun fitted to a trolley on a linear air track firing a table tennis ball. The toy gun moves in one direction while the table tennis ball moves in the other.

Apparatus required: •Linear air track •Toy gun •Trolley on an air track •Table tennis balls.

---

11   Stopping an escalator

I was walking down a moving escalator outside the Pompidou centre in Paris when a little boy hit the stop button. The next step, which I had assumed would be moving away, was then dead still and I received a severe shock to my back! This painful story is a good example of why you should always bend your knees on landing after any sort of jump and why you cannot make a jump without bending your knees first — time is needed to give you the required impulse (*Ft*). For a given change of momentum a small stopping time (as on the escalator) means a large force.

12 Pendulum on a trolley: momentum conservation

Hang a pendulum on a dynamics trolley or on the rider of a linear air track and observe its motion as the trolley moves along.

Apparatus required: •Linear air track •Pendulum.

13 Sand falling onto a top pan balance

The effect of the collision of a large number of particles such as gas molecules exerting a pressure or rain falling onto a roof can be simulated by using sand and a top pan balance. Put a plastic beaker on a top pan balance and pour a steady stream of sand into it from a known height. Record the reading of the balance while the sand is falling and also when no sand falls. Repeat the experiment with water falling into a beaker: both of these should show the effects of momentum change. Relate to the force and pressure on a roof in heavy rain (both flat and pitched).

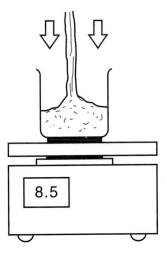

Theory: force on balance due to stationary sand $= mg$
Force on the balance due to a mass of sand m' that falls onto it every second with a velocity v and is brought to rest $= m'v$.

Apparatus required: •Sand •Plastic beaker •Top pan balance.

14 Crumpling zones in cars

This is a useful if rather unpleasant example of the energy and momentum conservation laws. I have some personal photos of this — our car! The bodywork crumples, absorbing kinetic energy and increasing both the stopping time and the stopping distance, so reducing the forces involved.

15 Collisions

The relative masses of the particles involved in a collision or explosion have an important bearing on what happens. Let's think of some elastic collisions.
    (a) If you throw a table tennis ball at a stationary train the train hardly moves and the ball will rebound with its velocity relative to the train reversed.
    (b) If the train is moving the train will move on with hardly any change to its velocity while the ball's velocity relative to the train will be reversed.

(c) If one ball hits a stationary ball of the same mass (as in snooker) the first ball comes to rest while the second moves off with the original velocity of the first.

(d) If one ball hits a stationary ball of rather greater mass (say ten times) the first ball will lose some of its kinetic energy, this energy being gained by the second ball.

This has an important bearing on the choice of moderator for a nuclear reactor. The neutrons must be slowed down by a moderator and those chosen have nuclei which are just a few times heavier than the neutrons themselves, such as deuterium or carbon.

16   Momentum in catching

This simple demonstration emphasizes the vector nature of momentum. Use a ball (soccer or netball) and throw it to a student. Get them first to catch it and throw it back and then to punch it backwards without catching it first. Ask them which require the biggest force — catching or punching it back. It should be clear that it is second and that this gives the biggest change of momentum. This demonstration hopefully emphasizes the velocity change from u to $-u$ and so a greater change of momentum than simply stopping the ball.

Theory: Momentum change $= mu - (- mu) = 2mu$ when it is punched back and momentum change $= mu$ when it is caught

Apparatus required: •Netball or football.

17   Helicopter details

The effects of momentum can be considered by a study of the motion of columns of fluids — air from helicopter rotors and water from a fire hose. It is useful to use the actual details of a Lynx helicopter to consider momentum transfer (mass = 4500 kg, rotor diameter 12.8 m). Force = $mg = \rho A v^2$ where ρ is the density of air, A is the area swept out by the rotors and v is the downward speed of the air. Other examples in this topic are fire hoses, moving walkways and the water from a propeller.

18   Throwing and jumping: momentum/inertia

Give various examples of the importance of impulse when you take a long time in kicking and catching. The longer the time during which the force can be applied the further you can throw something and the longer it takes you to bring the object to rest when you catch it by bringing back your hands the smaller the force needed. The effect of the forces involved in car crumpling on collisions should also be considered. Notice how long jumpers and high jumpers 'sink' on their last stride before take off to give time to apply an upward force.

In racket sports the same ideas apply. Also in cricket the ball can be hit further by maintaining contact between the ball and the bat by 'playing down the line' of the ball

19 Force in a long jump take-off or when a ball bounces

In experiment 18 it was suggested that estimates could be made of forces involved in jumping. This can be done by fixing contacts to the jumper's feet and to the ground. Using this method try to estimate the large forces generated as the jumper 'sinks' on the last stride. There is a very useful CD-ROM called *Multimedia Motion* which enables measurements to be taken from video clips. The time of contact between the foot and the ground can easily be found.

An alternative experiment is to find the force experienced by a ball when it bounces. This can be done by fixing a piece of aluminium foil to a ball and dropping it onto another piece so that it completes the circuit to a stop clock or scaler. Knowing the height from which it was dropped enables you to find the impact speed and so the momentum change and hence the force.

The contact area can be found by placing sand paper under a piece of aluminium foil so that when the ball is dropped on it an imprint is made in the foil.

Theory: force = change of momentum/time of contact = $(mv - mu)/t$

Apparatus required: •CD-ROM *Multimedia Motion* with computer.

20 Momentum and the rotating table

(a) Get someone to stand on the rotatable table holding a 1 kg mass. Ask them to throw the mass to you very carefully starting with their arm held out to the side of their body. As the mass moves forwards they will move backwards showing the vector nature of momentum.

(b) Stand a student on the table with their arms outstretched and push one of their arms gently so that the table and pupil spin slowly. Then ask then to bring their arms inwards. As they do so their rate of spin increases — the distribution of mass has changed and to maintain their angular momentum their spin rate must go up.

Apparatus required: •Rotating table •1 kg mass

FRICTION AND INERTIA

General background information for the friction section

Friction occurs between two surfaces in contact when they move relative to each other. Between two metal surfaces you often get 'stick–slip' motion because the surfaces weld together at a number of points on contact and these 'welds' break and reform as one of the objects is pulled over the other.

1. Lowering the friction
2. Styrocell beads
3. Stick–slip motion
4. Polystyrene ball falling in a tube
5. Two simple hovercraft

General background information for the inertia section

The inertia of an object can be described as its reluctance to change the way it is moving. Bodies with high inertia (large mass) are difficult to move and stop. The Apollo astronauts found their own inertia a problem when walking on the Moon. The lower gravitational pull meant lower friction between their feet and the ground so it was more difficult for them to overcome their inertia and stop. Heavy objects 'floating' in orbiting spacecraft are similarly difficult to control.

6. Coin/card and beaker
7. Inertia: car, tea set and blocks
8. Inertia
9. Inertia of a rod
10. Wig wag, or the inertia balance
11. Water on an umbrella
12. Rotational inertia: the bicycle wheel gyroscope
13. Inertia and the linear air track

Friction and air friction

1 Lowering friction with magnets

Take a bar magnet and mount it in a clamp with the poles vertically above each other. Take one or two clean ball bearings and attach them to the magnet so that they hang in a line from the lower end. Then attach a steel disc about 3 cm in diameter to the lower ball bearing. It should just be held on but not too tightly; if the attraction is too strong add another ball bearing. Blow on the side of the disc to set it rotating. Adjusting the number and size of the ball bearings and the mass of the disc will give a situation where the disc is only just supported and the friction between it and the lower ball bearing is then very low. Indeed, an old account of this experiment suggests that the disc could be made to rotate for over fifteen minutes.

Apparatus required: •Retort stand and clamp •Two or three ball bearings (diameter about 0.8 mm depending on magnet strength) •Strong bar magnet •Steel disc.

2 Styrocell beads

Friction, or the lack of it, can be demonstrated when a crystallizing dish is slid along over some stryrocell beads that have been poured into a tray. These beads are hard plastic spheres about a millimetre in diameter and so act like ball bearings. Spinning the dish makes a good demonstration and the whole experiment can be done in the base of a small ripple tank placed on an overhead projector to make it easily visible to the whole class. (Warning: don't let the students take any of the beads away — they are lethal on the floor!)

Apparatus required: •Styrocell beads •Ripple tank •Tray •Glass beaker.

3 Stick–slip motion

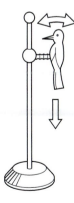

The stick–slip motion in friction is beautifully illustrated by the sliding woodpecker. This is a small wooden bird fixed to a bead by a short spring. The bead fits loosely over a vertical metal rod. When the bird is still it does not slide down but as soon as it is twanged it slips jerkily down the rod.

Theory/explanation: the hole in the wooden bead is slightly bigger than the diameter of the metal rod. When the hole is vertical the bead slips down but as soon as it tilts the bead 'grips' the rod. The oscillatory motion of the bird on the spring alternately puts the hole upright and at an angle.

Apparatus required: •Woodpecker toy (available from toy suppliers in England).

4 Polystyrene ball falling in a tube

Set up a vertical tube and drop a polystyrene ball into it. The diameter of the ball should be just a little smaller than the diameter of the tube. The ball falls really slowly due to air friction in the tube and it may be possible to get it to reach a terminal velocity where weight = drag!

Apparatus required: •Large diameter glass tube up to two metres long •Polystyrene balls of various diameters up to almost that of the internal diameter of the glass tube •Stop clock •Ruler.

5 Two simple hovercraft

(a) Balloon hovercraft. Drill a hole through the centre of a circular disc of hardboard (disc diameter about 10 cm). Place a rubber bung over the centre of the disc so that the hole in the bung coincides with the hole in the disc. Blow up a balloon and then fit it over the open end of the bung. Air rushes through the hole, and out the other side of the disc and the set up acts like a small hovercraft
  (b) Tile hovercraft. A simple hovercraft can be made really cheaply by using a computer cooling fan fixed above a hole in a ceiling tile.

Apparatus required: •Balloon • Hardboard disc with hole and tube •Ceiling tile •Computer cooling fan.

Inertia

6 Coin/card and beaker

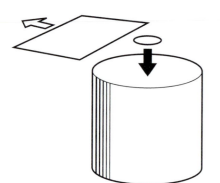

This is a simple demonstration of inertia. A card is put on the top of a beaker with coin on it. If you flick the card off sharply the coin falls into the beaker.
  The inertia of the coin makes it reluctant to move and the frictional forces between the coin and the card are too small to get it moving horizontally before it falls into the beaker.

Apparatus required: •Coin •Card •Beaker.

7 Inertia: car, tea set and blocks

Another two simple demonstrations of inertia.

(a) Get a friction-powered car running and then place it on a thin piece of cardboard that is lying on some styrocell beads. The car hardly moves while the card is moved backwards rapidly — an example of Newton's Second and Third Laws.

(b) Place a silk scarf on the bench and arrange a toy tea set on it. It is more impressive if some of the cups are full of water. Now pull the scarf out from underneath. The tea set should remain where it is. Trying it with light plastic toy cups and then with real ones demonstrates the effect of mass on inertia. Light cups have much less inertia and therefore move much more easily.

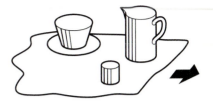

(c) Put a stack of three wooden blocks on the bench, one on top of the other. Give the middle block a sharp blow with a hammer. It should move out, allowing the upper block to drop onto the lower one, not moving horizontally much because of its inertia.

Apparatus required: •Friction-powered car •Card •Styrocell beads •Plastic tea set •China cup •Milk bottle.

8 Inertia

Tape two 1 kg masses together and then suspend them from a rigid support such as a beam by a piece of cotton and tie a second piece of cotton to the bottom mass. A sharp pull will break the lower piece of cotton while a gradual pull will break the top one. This demonstrates the effect of the inertia of the masses.

Theory: when the bottom thread is pulled gently the tension in it is T while that in the top thread is $T+Mg$ where M is the total mass of the two 1 kg masses. As the pull is gradually increased, clearly the top piece of cotton will reach its breaking stress first. However, if the lower thread is jerked sharply, the inertia of the masses prevents them moving before the cotton breaks. $Ft = Mu$ and if t is small then F is large.

Apparatus required: •Two 1 kg masses •Tape •Weak cotton, strong enough to support 2 kg but not much more •Rigid support such as a beam.

9 Inertia of a rod

The Victorians were fond of a parlour demonstration that used inertia. A needle was fixed into the two ends of a broom handle and then balanced with the two needles resting on wine glasses. The trick was to break the broom handle by hitting it in the centre with another broom handle without breaking the wine glasses. As a safer variation of this, hang a broom handle between two chairs by a piece of cotton at each end. Now either hit it in the centre with another broom handle or drop a heavy brick on the centre of the handle so that it will break without the threads breaking. The time of action of the force on the wood is short and it snaps before the force is transmitted along to the thread at both ends.

Apparatus required: •Two chairs or lab stools •Wooden rod (not too strong) •Brick •Carpet piece to protect the floor •Cotton.

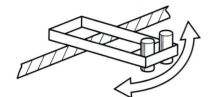

10 Wig wag, or the inertia balance

This is a classic experiment to compare the inertia of different masses. The apparatus is shown in the diagram. The tray fixed to two whippy metal strips is clamped to the bench and then displaced and allowed to oscillate; the frequency of oscillation can then be found. A 1 kg mass is now placed in one of the holes and the resulting frequency found. The procedure is repeated for up to three masses. As you would expect, the frequency goes down as the mass is increased, but is this due to the weight of the masses or their mass? An effective way of showing this is to repeat the experiment with one of the masses but suspend it by a thread so that it is hanging in the hole rather than resting on the base of the tray.

You should find that the oscillation period is unaltered, showing that it is the mass that governs the vibration rate and not the weight. In other words, the frequency would be the same in an orbiting spacecraft or on the surface of the Moon as it were on Earth. This makes it a very useful as a timing device for astronauts.

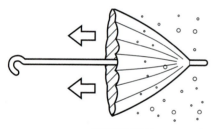

11 Water on an umbrella

This is a simple demonstration of inertia. Simply get an umbrella and sprinkle some water on it; you need a really waterproof surface to make this work well. (If your umbrella is old try and get it reproofed or else smear the surface with a little oil.) Now close the umbrella sharply; the water stays where it is due to inertia and so the umbrella comes away much drier. This is rather like a stationary car being hit from behind. The passengers remain where they are and the car moves off (see experiment 13).

12 Rotational inertia: the bicycle wheel gyroscope

There are some very impressive demonstrations of the inertia of rotating objects. One of the most simple and yet effective methods is to use a large gyroscope such as a small-sized bicycle wheel that has had two wooden handles fitted to either side of the axle.

Get someone to hold the handles, wind a length of string round the axle, and then, using a strong pull, set the wheel spinning. Then ask them to hold it by just one handle and then change the direction in which the axle is pointing. It is very difficult because of the torque needed to alter the angular momentum of the wheel.

A number of fascinating experiments can be performed with this heavy gyroscope:

(a) get the students to try to pass it round the class while it is rotating using only one hand each and then ask them to abruptly change the axis of rotation;

(b) get them to stand on a rotating table, hand them the spinning wheel with its axis horizontal and then ask them to turn the axis vertical;

(c) suspend the spinning wheel by only one side of its axle — precession and nutation will result;

(d) mount this spinning wheel in a case and get a student to carry it round the lab.

Apparatus required: •Rotating table •Bicycle wheel gyroscope •Length of cord.

---

13 Inertia and the linear air track

An interesting extension to experiment 7 is to use a linear air track mounted on wheels! I use two old dynamics trolleys. Use a reasonably heavy rider (say 400 g) and a two metre long linear air track. Put the rider on the track, turn up the air supply so that it floats and then pull the track itself — I use a weight hanging over a pulley. The inertia of the rider and the very low friction between it and the track will keep it motionless while the track moves under it. Make sure that you stop the air track before it falls off the bench and have somebody control the air hose!

Apparatus required: •Linear air track •Rider •Pulley •Two small wheeled trolleys such as dynamics trolleys •Air supply •Thread •Mass.

VECTORS, MOMENTS AND STABILITY

General theory for this section

This section deals with the moment of a force, balancing, centre of gravity and stability.

The moment of a force is defined as the product of the force and the perpendicular distance from the line of action of the force to the pivot.
When an object is balanced the sum of the clockwise moments (those trying to turn it in a clockwise direction) is equal to the sum of the anticlockwise moments (those trying to turn it the opposite way).

When an object is balanced the vertical line from the centre of gravity passes through the point of balance and the object will topple over when this vertical line falls outside the base of the object.

1. Balancing forks and a pivot mechanism
2. Balancing a pencil and/or a snooker cue
3. Pile of leaning blocks
4. Shopping trolley
5. Mop and back muscles
6. Action Men and artists' models in physics
7. Interesting balancing
8. Male and female balancing
9. Mobiles to demonstrate moments
10. Arm muscles and levers
11. Centre of gravity of a student
12. Moments
13. The centre of gravity, the wooden spoon and the mass of a broom
14. Rolling up
15. Moving fingers on a long ruler
16. Arm wrestling
17. Vectors
18. The heavy bottom toy

1   Balancing forks and a pivot mechanism

(a) The position of the centre of gravity of a system is vital for its stability. For an object to be in stable equilibrium its centre of gravity must be below the point of suspension. This can be demonstrated clearly by sticking two forks into the opposite sides of the cork, fixing a coin vertically into the lower base of the cork and then balancing the coin on the tip of a pencil.

(b) A simple way of lowering the centre of gravity of a ruler when used for investigation of the law of moments is to fix a rubber band round a rod to be used as the pivot with the ruler below it balanced on an aluminium yoke. This gets the centre of gravity below the point of suspension and makes balancing easier.

Apparatus required: •Two forks •Cork •Glass •Ruler •Rubber band •Metal rod.

2   Balancing a pencil and/or a snooker cue

Why is it so much more difficult to balance a pencil than a snooker cue on your finger? This can be demonstrated by using a number of different shaped rods — long, short, light, heavy with a mass at the top or near the lower end, a pencil and a snooker cue. It is much easier to balance heavy rods or ones with a mass at the top such as a mop!

The ease of balancing is simply to do with how sensitively you can adjust the position of the centre of a mass; a small movement of your hand will produce a large angular acceleration in a light rod.

Apparatus required: •Rods of different types: snooker cue, broom, pencil, metre ruler, six inch nail.

3   Pile of leaning blocks

The idea of this demonstration is to look at stability. Take some rectangular blocks (dominoes are ideal) and pile one on top of the other setting each domino a little off-centre compared with the one below it. How many can you pile up and where is the centre of mass of the whole stack?

Theory: a mathematical rule can be found to give the greatest overhang for a given number of blocks. It can be shown that the Nth block can have a maximum projection of $1/(2N)$ relative to the block below. In other words the top, or first block can hang $1/2$ its length over the second block, while the second block can hang $1/4$ of its length over the third and so on. For a large number (N) of blocks the sum of the series is $0.5(0.5772 + \ln N)$.

Apparatus required: •A number of dominoes or other suitable blocks •Ruler.

4 Shopping trolley

This is an interesting example of everyday balance. The trolleys are extremely stable and make a good example for the discussion of moments; the handles are almost directly above the back wheels. Have you ever tried to tip one over from the back? Even if you put your full weight on the handles the trolley will not tip over. (It is useful if you can actually borrow one and bring it into the lab!)

Apparatus required: •Shopping trolley.

5 Mop and back muscles

The following demonstration is a simulation of the enormous tension produced in the muscles of your back when you lean over. Tie a piece of string to the handle of a mop about a quarter of the way from the mop head. The head of the mop represents your head and the handle of the mop represents your spine. Drill a hole through the end of the handle furthest from the head and pivot it here, the head of the mop being at the top.

Now try and support the mop as it tilts by holding the string at a small angle to the handle of the mop (I am told that the muscles make an angle of only 10° with your spine!) The tension in the string represents the huge tension in your back muscles as you bend over. Bending at 45° produces a tension of over double your own body weight!

Holding something in your hands while you bend over (e.g. into a car boot) will increase the tension even further.

Theory: the tension in string (back muscles) T is given by the equation:
$T \sin A = mg \cos \theta$
where A is the angle of the string with the mop (back muscles with the spine) and θ is the angle that the mop handle (spine) makes with the horizontal.

Apparatus required: •Mop •String •Newton meter •Metal rod for pivot •Retort stand and clamp •G-clamp.

6 Action Men and artists' models in physics

The stability of the human body and the variable position of its centre of gravity can be studied using wooden artists' models, Action Men or even Barbie dolls. Their ability to stand, bend and balance on the flat, on trolleys and even on rotating tables is much easier to investigate than using real people who might get hurt!

Apparatus required: •Action Man •Small wooden artists' model.

7   Interesting balancing

(a) Balance two beakers of water on a lever balance and then put your finger in one of them. What happens? You can do this by using one beaker and a top pan balance (see the section on Archimedes and upthrust).

   (b) Put two pieces of taper in the end of a straw and, using a pin as a pivot, balance the arrangement. Now light the tapers — what happens? This can also be done with a candle which as been sharpened at top and bottom so that it can be lit at both ends while being pivoted in the middle. A fascinating rocking motion results.

Apparatus required: •Lever balance •Beaker of water •Taper •Straw •Candle •Pin and cork •Retort stand and clamp.

8   Male and female balancing

An interesting and amusing experiment on the balancing of the human body that can be done with your older students is to get first a male student and secondly a female to stand facing a wall, three foot lengths away from it and with their hands behind their backs. Now ask them to bend forwards so that their nose just touches the wall. Female students are supposed to be able to do this without overbalancing while male students are not. Why is this?

9   Mobiles to demonstrate moments

A lot of simple physics can be demonstrated with a child's mobile. It is instructive to design one both by trial and error and also by working out the moments required at each level. The ones with the more massive hanging objects tend to be better since they are more stable. I have seen a lovely one in France at a zoo in the Loire valley made with some two dozen small wooden fish hanging by threads from wire supports.

Apparatus required: •Objects to construct a mobile •Cardboard •Balance •Stiff wire •Thread.

10   Arm muscles and levers

A study of lifting weights by your arm is a good example of levers. The muscles of the forearm have a much greater tension in them than the weight you are lifting if the forearm is held horizontally. Get one of the pupils to try this by first holding the weight near their shoulder and then slowly extending their arm — the torque (turning effect) becomes greater.

Theory: the moment of a couple or torque = force × perpendicular distance between the line of action of the two forces.

Apparatus required: •Masses (up to 5 kg is useful) •Model forearm if possible.

---

11 Centre of gravity of a student

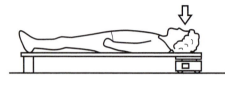

You can find the centre of gravity of a student by the following method using a strong wooden plank, a pair of bathroom scales and a brick (or block of wood the same height as the scales). Put the plank down with one end on the block of wood or brick and the other on the bathroom scales. Lie the student on the plank with their heels over the pivot (brick). Record the reading of the scales. Take moments about the brick having weighed the student first.

Theory: reading on scales × distance of scales from pivot = weight of student × distance of student's centre of gravity from the pivot + weight of plank × distance of centre of gravity of plank from the pivot.

Actually the weight of the plank can be ignored if you record the increase in the scale reading when the student lies down.

Apparatus required: •Plank (2m long) •Bathroom scales •Metre rule.

---

12 Moments

A very simple demonstration of the effect of distance on the turning effect of a force can be demonstrated using a door. Get a small student to close the door by using with one finger placed near the edge of the door while the teacher pushes with full strength on one hand very close to the hinge! The effect of the distance of a force from the pivot is clear — the student usually manages to push the door closed!

An extension of this for older students is to get one of them to push the door such that their arm is at an acute angle with the door. It is much harder to open or close it. This emphasizes the importance of 'perpendicular distance' in the definition of the moment of a force.

---

13 The centre of gravity, the wooden spoon and the mass of a broom

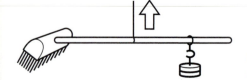

This simple experiment emphasizes that it is not just the mass on either side of a balance point that determines whether an object will be balanced but also how it is distributed. First demonstrate this with a wooden spoon. Balance it on your finger to find the centre of mass and then cut it into two pieces through the centre of mass and weigh the two pieces showing that their masses are not equal. Then go on to the main experiment.

Hang up a broom from roughly the centre of its handle so that the head will go down, the centre of mass of the broom being on the side nearest the broom head. Now by loading the other end with masses (m) bring the broom into a horizontal position. Locate the centre of gravity of the broom by removing the masses and adjusting the position of the string so that the broom balances.

Theory: force of weights × distance of weights from pivot = weight of broom × distance of centre of mass from pivot.

Apparatus required: •Retort stand and clamp •String •Broom •Large wooden spoon • Slotted masses.

14   Rolling up hill

An interesting demonstration of the turning effect of a force is to make a tin roll up hill! Use a large flattish cake or biscuit tin (diameter 30 cm and depth 8 cm or so works well) and fix a hidden mass (such as a lump of plasticine) inside one rim and then show the pupils that the tin can roll uphill. Of course you will start it so that the mass is slightly to the up hill side of the vertical and the tin will only roll up hill until the mass reaches the lowest point — hence the advantage of a fairly large tin. See how long it takes them to spot what is going on without getting hold of the tin!

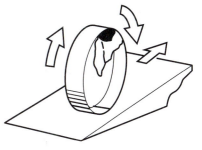

An alternative method is to use two large discs cut from thick sheets of polystyrene. A depression is made near their rims and a lump of modelling clay wedged in it. The two discs are then stuck together, the loaded section being invisible!

Apparatus required: •Plank to act as a ramp •Large tin with lid and lump of modelling clay.

15   Moving fingers on a long ruler

This is a surprising demonstration of the effect of frictional forces and balancing. Balance a meter ruler or a broom handle on two fingers, the fingers being at different distances from the two ends. Then slide your fingers together; they will always meet in the centre of the ruler as long as it is uniform. This is good for a discussion of moments. Of course, if the rod is not uniform, such as a snooker cue, the two fingers will meet at the centre of mass. (See the example about the polar bear on ice, miscellaneous mechanics no. 26.)

Theory: the greater the distance of a finger from the centre of mass of the ruler the smaller the reaction on it and so the smaller the frictional force.

The ruler will then slide more easily over this finger than the other which is nearer the centre of mass. This situation continues until they are at equal distances, when both fingers move together.

Apparatus required: •Metre ruler or other long uniform rod.

---

16 Arm wrestling

An example of balancing two forces and/or resultant forces. The length of arms and the turning effect can also be mentioned here.

---

17 Vectors

A very simple introduction to the importance of the directional nature of a force can be given by using a rope held horizontally and pulled tight by two strong students, one pulling on either end. Now a relatively weak student should try to push down on the centre of the rope; it is quite easy to move it down no matter how hard the other two pull.

Theory: downward force $= 2T \sin A$ where T is the tension in the rope and A is the angle that the rope makes with the horizontal. Say $T = 200$ N and $A = 10°$, i.e. a 2 m rope pushed down in its centre by about 15 cm. This requires a downward force at the centre of only 34 N

Apparatus required: •About two metres of rope •Newton meter as an extension.

---

18 The heavy bottom toy

This toy can be used to demonstrate stability. It has a low centre of gravity and will always return to the vertical position if displaced.

CIRCULAR MOTION

General theory for this section

Centripetal force is the force that pulls or pushes an object from its straight-line path. It always acts towards the centre of a circle. The centrifugal force is the reaction of this force on the thing doing the pushing or pulling but not on the rotating object itself.

Centripetal force $= mv^2/R$ where m is the mass of an object moving at a constant speed v round a circle of radius R

1. Model prop powered plane on a thread: circular motion
2. Shape of rotating liquid surface
3. Whirling bucket
4. Rotary water sprinkler
5. Rotating lawn sprinkler
6. Toy cars and loop the loop
7. Wall of death: fruit bowl and Marmite lid
8. Fairground rides
9. Rounders bat on rotating table
10. The wire coat hanger and circular motion
11. Rotating candle: flame bends inwards
12. Back seat of a car
13. A simple centrifuge by whirling a container on string
14. Rotating jelly:
15. Rotation of rigid bodies and moments of inertia

1 Model prop powered plane on a thread: circular motion

The theory of a conical pendulum can be very clearly demonstrated by using a model battery powered plane that hangs from a pivot by a thread, the pivot being able to rotate. The battery drives a large propellor at the back of the plane making it fly in a circle of radius r at a constant speed — a good simulation of the chairs in fairground rides. The faster the plane the bigger the angle (θ) that the string makes with the vertical. Measurements of θ, v, m and r are easy to make. The only problem — it's difficult to stop!

Theory: for a plane of mass m, resolving vertically $mg = T \cos \theta$; resolving horizontally $mv^2/r = T \sin \theta$.
Therefore $\tan \theta = v^2/rg$

Apparatus required: •Model plane on thread and suitable support •Stop clock •Ruler

2 Shape of rotating liquid surface

The shape of a rotating liquid surface can be found using the following experiment. Mount a glass beaker securely at the centre of a rotating table. (One of the best methods is to make a hollow circular aperture in a piece of wood into which the beaker will just fit and screw or bolt this to the rotating table). Put warm melted wax in the beaker and spin it. As the wax cools and solidifies a permanent record of the shape of the rotating surface will be produced. Using water, oil or wallpaper paste will give a temporary record.

An alternative version is to use a bowl of sugar. The shape of the surface can be retained as long as you don't spin it too fast!

Theory: shape of the surface can be shown to be $y = \omega^2 x^2/2g + C$ where ω is the angular velocity and x the distance from the centre of rotation.

Apparatus required: •Wax •Beaker •Rotating table •Motor and drive belt •12V variable DC supply.

3 Whirling bucket

The classic centripetal force experiment. Put a little water in a bucket, tie a string firmly to the bucket handle and then swing the bucket in a vertical circle. As long as the rate of rotation is great enough the water stays in the bucket! Slowing the rate of rotation can get the water to almost fall out at the top of the path and you can usually hear it slopping around at this critical point. Mention that the water and bucket experience a centripetal force but that there is also a centrifugal force — this is the reaction on the pivot, in this case your hand. Make sure that the handles of the bucket do not come off and that the bucket does not hit the floor at the lowest point of the circle. Many extensions of this are possible such as swinging a tray loaded with beakers by four strings!

Theory: the water and the bucket move in a vertical circle at a constant speed and so although the centripetal force is constant the tension (T) in the string varies. It is greatest at the lowest point of the circle because of the differing contributions of gravity at different points of the circle.

Centripetal force $= mv^2/r = T + mg\cos\theta$ where θ is the angle that the string makes with the vertical measured from the point where the bucket is at the top.

Apparatus required: •Bucket •String •Water.

4 Rotary water sprinkler

Fix an inflated balloon to the water inlet nozzle of a rotary lawn sprinkler and allow the balloon to deflate. Use this not only to show circular motion but also to measure energy conversion. I prefer to block up the vertical outlet, if the sprinkler has one, so that air only emerges horizontally

Apparatus required: •Rotary water sprinkler •Balloon •Stop clock to measure time of rotation.

5 Rotating lawn sprinkler

Use the lawn sprinkler again but this time fixed to a water tap. This is a good example of angular momentum. The momentum of the water leaving the sprinkler imparts an equal and opposite momentum to the central rotating head.

6 Toy cars and loop the loop

Toy cars on a plastic track can be used to demonstrate a number of ideas in mechanics. If the track can be bent into a vertical circle then loop the loop experiments can be performed.

Theory: centripetal force $= mv^2/r$ This is provided by a combination of the reaction of the track (R) and the force of gravity (mg). At the top of the path the car can travel at its minimum speed since the centripetal force can be provided by gravity alone. $mv^2/r = R + mg$. The greatest reaction occurs at the bottom of the path where $R = mv^2/r + mg$

Apparatus required: •Toy cars •Loop the loop track •Ruler.

7 Wall of Death: fruit bowl and Marmite lid

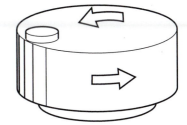

This demonstration is a simulation of a motorcyclist on a Wall of Death at a fairground. Take a glass or plastic fruit bowl of diameter at least 30 cm to represent the Wall of Death and a lid of a jar to represent the motorcyclist (one from a Marmite jar works well). With a little practice the lid can be made to roll round the sides of the bowl on its edge and once moving can be kept going by a small oscillation of the bowl. It will roll round rapidly even though the sides of the bowl are kept vertical.

Apparatus required: •Fruit bowl and Marmite lid.

8 Fairground rides

(a) Ride one — loop the loop

Using the loop the loop track with toy cars is a very good simulation of what happens in vertical fairground rides. The question is: how high up must you start the ride so that the cars just make the loop without falling off? It is difficult to demonstrate this exactly because of friction but we can get an idea. Theoretically the car must start from a height which is 2.5 times the radius of the loop.

Theory: at the top of the loop $a = v^2/r = g$ if the car is not to fall off. Therefore kinetic energy at this height must be $\frac{1}{2}mv^2 = \frac{1}{2}mrg$. Potential energy $= mg^2r$ and so the total energy at the top of the loop is $2.5mrg$. So if the car is to make the loop without falling off it must have had an initial potential energy of $2.5mrg$ and so must begin from a height of $2.5r$ above the base of the loop.

(b) Ride two — the swinging chairs

These are chairs fixed to a central pillar by wires. As the rate of rotation is increased, the chairs swing out further and further from the vertical. Do they all swing out the same distance?

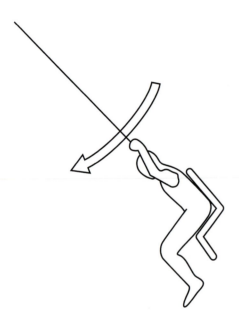

You can make a simple model of the chair ride at a fair with chairs made from bottle tops. Mount it in the centre of a rotating table and then spin the table using a motor. It can be used to demonstrate the effect of different masses in the chairs by loading some with plasticine, you could make actual 'people' shapes. The important thing is that the mass does not affect the angle of the strings to the vertical.

Theory: consideration of the formula $\tan\theta = v^2/rg = r\omega^2/g$ (where v is the linear velocity and ω the angular velocity of the chairs) shows that all the chairs swing out the same amount regardless of their mass.
θ is the angle that the string makes with the vertical, v the speed of rotation and r the radius of the circle.

Apparatus required: •Chairs model •Motor •Suitable power supply •Rotating table •G-clamps.

9 Rounders bat on rotating table

Stand on a rotating table and swing a rounders bat to show the effect of conservation of angular momentum. This is a more effective extension of trying to rotate yourself on a rotating table where moving your arms in one direction causes your body to move in the other, both coming to rest at the same time. Standing on the table and very carefully throwing a kilogram mass is another effective demonstration. Be careful about how and when the mass is thrown!

Apparatus required: •Rounders bat •Rotating table •1 kg mass.

10 The wire coat hanger and circular motion

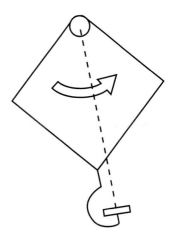

Pull open a wire coat hanger so that it forms a square. File the end of the hook flat and then bend the hook until it points towards the opposite corner of the square. Balance a 1p coin on the hook, put one finger in the corner of the square opposite the hook and then spin the coat hanger in a vertical circle — the coin stays in place! This is a very simple but excellent demonstration of centripetal force.

The force of the hook on the coin always acts towards the centre of rotation. The current record is five 1p pieces stacked on top of each other. With only one penny balanced and with great care I have once even been able to bring the coat hanger to rest without the penny falling off.

Apparatus required: •Wire coat hanger with filed end •Pile of small coins.

11 Rotating candle: flame bends inwards

Put a candle on a turntable shielded by a glass tube such as a jam jar; the candle must be shorter than the height of the jar. Light the candle and rotate the turntable. Watch the flame! (Diameter of table 30 cm and a rotation rate of about 1 Hz is appropriate.)

Apparatus required: •Rotating table •Motor •Jam jar •Candle •Blu tac.

12 Back seat of a car

Who falls into whose lap as you go round corners? The person nearest the centre of the curve travels on in a straight line whereas the one on the outside of the curve is pushed round by the side of the car to meet them. So it appears that the person on the inside falls into the other one's lap. This is similar to the effect on the clothes in a spin drier. Notice that if the driver overdoes it the car will roll outwards, the inner wheels leaving the ground first.

13 A simple centrifuge by whirling a container on string

A simple centrifuge can be made by whirling a large test tube round your head on a piece of string. Use a mixture of water and sand to show the separation. Experimenting with other liquids such as syrup and wallpaper paste makes an interesting extension to this.

Theory: $F = mv^2/r$.

Apparatus required: •Test tube •String •Water •Wallpaper paste •Sand

---

14   Rotating jelly: circular motion

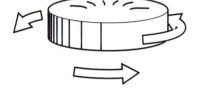

The effects of centripetal forces on a rotating object can be shown impressively by making a circular jelly about 3 cm deep in a crystallizing dish. When it is set, empty it out carefully onto the centre of a saucer which is securely fixed to the centre of the rotating table (use a safety screen!). Slowly increase the rate of spin of the table. The jelly will flatten.

Further increasing the rate of spin will eventually make the jelly break up — the cohesive forces within it being less than the centripetal forces needed. It can be used to demonstrate why car tyres fly apart when they are spun too fast. (I was told that if you used some of the old forms of tyre remoulds that you should not travel at more than 60 mph to reduce the risk of the tyres breaking up!)

The jelly experiment also shows the shape of the liquid surface while rotating. It is useful to photograph it or take a video for later analysis.

Apparatus required: •Crystallizing dish •Jelly •Saucer fixed to rotating table •Power supply •Motor

---

15   Rotation of rigid bodies and moments of inertia

This topic can be demonstrated clearly by using:

(a) discs of equal mass but of different mass distribution as energy sources;
(b) inertia powered toy cars;
(c) a heavy bicycle wheel.

ELASTICITY

General theory for this section:

An object obeys Hooke's law if the deformation is directly proportional to the applied force. For example this will apply to a piece of copper wire being stretched by small loads; the actual maximum load depends on the cross sectional area of the wire.

The Young modulus (E) for a material is a measure of how much it will stretch. This is given by the equation $E=FL/eA$ where e is the extension, L the original length, F the applied force and A the cross-sectional area of the wire. For steel the Young modulus is 2×10^{11} Pa.

1. Uses of elasticity
2. Bending of a beam
3. The belly flop
4. Glass
5. Energy stored in a rubber band
6. Shear stress
7. Silly putty
8. Bungee jumping
9. Stretching of a sock or tights
10. Strength of paper and string
11. Elasticity of rubber molecules
12. Heating in stretched rubber
13. Creatures and granite columns: leg strength, height and g
14. Whirling springs and Hooke's law
15. Elasticity of rubber when cooled
16. Electric strain gauge

1 Uses of elasticity

The following is a list of the applications of elasticity that I have found useful when introducing the topic: a bouncy castle, space hopper, elastic in clothes, bridges, trampolines, training shoes, cricket bats, aircraft wings, muscles, skin, car bumpers, crash helmets, car and bike suspension and fishing lines. Some mountain bikes actually have air suspension forks!

2 Bending of a beam

As a practical example of bending and shear stresses, bend a metre ruler in the laboratory by supporting it at either end and then loading it at its centre. Investigate the effect on the depression of changing the load and the length of the beam.

A rather nice extension of this uses a filament of glass drawn out from a glass rod, the filament being in the centre of two much fatter end sections. The rod should be held horizontally by its end and then the filament in the centre loaded with masses. A quite surprising deformation can be produced. (Do this behind a protective screen and wear goggles to protect your eyes in case the glass fractures!)

Theory: The depression of the centre of the beam when it is loaded at its centre is proportional to the load and to the length of the beam cubed.

Apparatus required: •Slotted masses •String •Metre rule •Two knife edges on bosses •Retort stands •Additional ruler mounted vertically in base clamp or TV camera and scale •Glass rod with central filament section •Safety screen •Goggles.

3 The belly flop

An idea of the very high compressibility modulus of water can be gained from talking about how painful it is to do a 'belly flop' into a swimming pool. It's actually only 100 times worse to launch yourself onto a slab of concrete if the water does not part to let you enter!

Theory: bulk modulus of concrete $= 100 \times 10^{11}$ Pa;
bulk modulus of water $= 1 \times 10^{11}$ Pa

4 Glass

The elastic nature of glass can be seen in old windows. The weight of the glass makes it 'creep' — the lower portion of the pane becoming fatter than that above it. Glass is much easier to break when it is scratched; the scratch provides an initial dislocation which then spreads easily when under stress.

Have you ever wondered why it is that a glass fibre may be bent in a circle while a glass block could only suffer small angular deformation before it shatters? This is because of the difference in length between the two sides of the specimen — small for the fibre but large for the block for a given angle of deformation.

5 Energy stored in a rubber band

You may know that if a rubber band is stretched considerably beyond its elastic limit it will not return to its original length. This can be shown by adding masses to the lower end of the band and then removing them carefully, recording the corresponding extensions. I have found it better to cut the band into one straight length giving greater extension for a given mass.

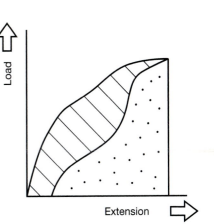

The non-recoverable energy absorbed by the rubber band may be worked out by finding the area of the hysteresis loop produced by the force–extension graph. It is instructive to calculate the velocity of projection of a paper pellet by a rubber band both with and without allowance for the energy that cannot be recovered. Also can air resistance be ignored?

A very rough estimate of the energy stored in a rubber band can be gained by firing a ball upward from a rubber band stretched to a known amount. Of course air resistance cannot be allowed for here.

Theory: non-recoverable energy absorbed by the band = force × distance = area inside the hysteresis loop.

Apparatus required: •Rubber band •Slotted masses •Ruler •Base clamp to hold the ruler vertical from the floor •Retort stand and clamp •Suitable clamp for holding the rubber band.

6 Shear stress

This can be demonstrated in a number of ways and it should be pointed out that it is this type of stress that is most likely to be encountered in machines and buildings.

(a) You can show it clearly with a simple pack of playing cards. If the top card is slid sideways while being pressed down on the cards beneath the whole pack will exhibit shear. Don't use cards that are too slippery!

(b) Make a 'sausage' of modelling clay and twist it at one end while holding the other end steady.

(c) Clamp a metre rule to the desk and twist that.

(d) Show how you can rotate your arm, mainly the wrist.

Apparatus required: •Pack of playing cards •Modelling clay •Ruler.

7 Silly putty

This is wonderful stuff and is available in many toy shops but if you see some, buy it! Stocks may not last. There are two different types:

(a) the type that glows in the dark

(b) the type that changes colour when you heat it.

Both types have the same elastic properties. If a ball of it is held by one side it will creep (slowly sag), if dropped it will bounce, and if hit with a hammer it will shatter.

Apparatus required: •Silly putty •Hammer •Ultraviolet lamp •Safety spectacles •Ultraviolet goggles

---

8   Bungee jumping

This can be simulated by using a long piece of elastic and a mass, doing the experiment over a stairwell if possible. Elastic bought from a haberdasher's works well or you might try unravelling a piece of climbing rope. Can you get the length right so that the mass stops just before it hits the ground?

Apparatus required: •Long length of elastic •Masses.

---

9   Stretching of a sock or tights

This experiment relates elasticity to something practical — clothes!

Take some tights (one leg will do) and fix them to a clamp. Load the bottom end with weights and record the force and resulting extension. (Forces of up to 150 N are needed to given a reasonable extension.) Slowly remove the weights and record the extension as they are taken off. Tights are very good; they stretch a lot when loaded and also show a hysteresis effect which apparently disappears when they are washed!

Apparatus required: •Tights •Masses •Retort stand, boss and clamp •Ruler.

---

10   Strength of paper and string

This is an interesting set of experiments to demonstrate the difference between the shape of a specimen and the way forces are applied to it. Try and stand a sheet of paper on one end with a 100 g mass on top — the paper collapses. Now roll the paper into a tube — it will easily support the weight of the 100 g.

Hold a piece of string vertically with the 10g mass tied to the top; let go and the mass falls as the string collapses. Now hold the string from the other end so that the mass is at the bottom; the string easily supports it.

Apparatus required: •Paper •100 g mass •String.

11 Elasticity of rubber molecules

Stretch a piece of rubber until it becomes difficult to continue. At this point the rubber molecules should have untangled themselves and should be aligned along the direction of stress. You will now probably notice that the band looks a little paler and the surface texture may have changed. Now put a pin through the rubber and move it about a little. The rubber will break apart along the line of stretch, which is also along the line of molecules showing that there is a lower strength between molecular chains than along them.

Apparatus required: •Two clamps •Retort stands etc •Large slotted masses (up to 5 kg) •Pin •TV camera if possible.

12 Heating in stretched rubber

Put a piece of balloon rubber between your lips and stretch it. As it is stretched you will notice an increase in temperature unlike what happens with a gas, where expansion usually means cooling. Alternatively, use a rubber band about 1 cm wide; holding a length of a centimetre or so between your thumbs and forefingers pull it sharply, then place it quickly against your top lip. The heating effect is very noticeable. Is it that the molecules in the rubber have become more ordered and so they give off heat energy, resulting in a cooling?

Apparatus required: •Rubber balloon.

13 Creatures and granite columns: leg strength, height and g

(a) We are made the way we are because of the gravitational pull of our planet. What would the design of other creatures be on planets with widely differing g values? Animals living on planets with very high g values would be expected to have thick legs while those of similar mass that lived on small planets with low g values could be supported by spindly legs!

While we are thinking about creature design, what about considering if animals would black out if they had a long neck and turned round too quickly.

Bone is actually amazingly strong; the compressive breaking stress of solid bone is only about a third that of steel! (170×10^6 Pa). A piece of chicken bone can be investigated to show this strength. Safety goggles and a safety screen are essential to protect you from flying splinters. The bone is held vertically in a clamp and weights placed on the top of a small platform resting on the bone.

(b) The maximum possible height of a granite column on Earth is around 7800 m. Higher than this and it will shatter under its own weight.

Simulate this using a column of damp sand. Put a tin lid on top and load this with masses until the sand column gives way. This should occur at the base. Another way of doing this is to use jelly. Make a tall jelly in either a gas jar, measuring cylinder or glass tube. Pour it out vertically and see how much can be outside the support of the tube before it splits apart.

Theory: breaking stress $= h\rho g$ where h is the height of the column and ρ its density.

Apparatus required: •Sand •Water •Masses •Tin lid •Chicken bone •Jelly •Safety goggles and safety screen.

---

14   Whirling springs and Hooke's Law

Put a small mass on the end of a spring and spin it in either the vertical or horizontal plane. To measure the force you can use a spring balance in the line with a string. Alternatively, calculate the value by measuring the radius of orbit and the speed of rotation. Record the reading on a TV camera if possible for slow analysis later.

Apparatus required: •Helical spring •Mass •Ruler •Stop clock •Spring balance •Optional TV camera.

---

15   Elasticity of rubber when cooled

Many of us will have seen, or heard, about the wonderful demonstrations of the change of elasticity of rubber when cooled in liquid nitrogen. However, they also work using dry ice (solid carbon dioxide). Try cooling a piece of rubber tube in solid carbon dioxide for a minute or two and then hitting it with a hammer; the rubber shows a marked change in its elasticity and will shatter under the impact. The change in elasticity of a squash ball and a rose petal when put in the solid carbon dioxide is also worth investigating.

Apparatus required: •Squash ball •Rubber tubing •Flower petal •Hammer •Carbon dioxide cylinder and cloth.

---

16   Electric strain gauge

A simple model of an electric strain gauge can be made by an arrangement similar to that used for investigating the stretching of fishing line. Take a 2 m length of resistance wire (such as nichrome) and clamp one end to a bench. Pass the other end over a pulley and hang a weight on the end. Measure the resistance between two points on the wire as far apart as possible either by using a sensitive ohmmeter or a voltmeter and a microammeter. (Do keep the current low — you do not want any heating effect). Now increase the load and a change in the resistance of the wire should result.

Apparatus required: •Nichrome wire (2 m) •Set of weights •G clamp •Bench pulley •Ruler •Ohmmeter or voltmeter and microammeter.

FLUID FLOW AND VISCOSITY

General theory for this section
There are basically two principles involved here:

(a) Bernoulli's law, which relates to the lowering of pressure in a moving fluid;

(b) Stokes' law, which relates to the terminal velocity of objects falling through a fluid.

The viscous drag on a sphere of radius r falling at its terminal velocity (v) through a fluid of viscosity $\eta = 6\pi\eta rv$

1. Fluid flow: ink and glycerol
2. Stokes' law
3. Fluid flow in tubes
4. Balancing balls
5. Viscosity and falling dust
6. Styrocell beads in a snowstorm
7. Throwing a ruler (Bernoulli)
8. Blowing between two sheets of paper
9. Moving a ball in the air
10. Ball in funnel

1 Fluid flow: ink and glycerol

Fill a measuring cylinder or gas jar with glycerol (wallpaper paste might do) and put a thin layer (a few mm) of blue ink on the top. Drop a ball bearing into the jar. As it falls some of the ink is collected by it and the streamlines round the falling ball are clearly visible as it moves downwards. With care the ink can be pipetted up again, so making the glycerol reusable!

Apparatus required: •Tall measuring jar •Glycerol or wallpaper paste •Ink •Pipette •Ball bearing.

2 Stokes' law

The effect of viscous drag on a sphere can be investigated by dropping ball bearings through glycerol or wallpaper paste. Alternatively you can watch bubbles rising through a fizzy drink, syrup or glycerol.

Theory: when the balls (radius r) are falling at their terminal velocity (v) through a fluid of coefficient of viscosity η the drag due to the viscous effect of the fluid is equal to the difference between their weight (mg) and the upthrust (U) $mg - U = 6\pi\eta rv$

Apparatus required: •Tall measuring cylinder •Glycerol •Wallpaper paste •Stop clock.

3 Fluid flow in tubes

Compare the viscosities of a number of different liquids by allowing them to flow down through glass tubes. It has been suggested that you could make a clock from a tin of syrup — graduated by how long it takes the syrup to run out!

Apparatus required: •Various liquids: water, syrup, glycerol, wallpaper paste •Tin with a hole in the bottom •Glass tubes of various diameters •Stop clock •Filter funnel •Rubber bung with a hole in the centre for attaching the glass tube to the filter funnel •Short length of rubber tubing •Tube clip.

4 Balancing balls

Balance a polystyrene ball or a table tennis ball on top of a vertical jet of water from the tap or a vertical jet of air from an air blower (the first one is messier!) With the air jet it is possible to bend the jet at quite an angle with the vertical, showing that it is not simply the upward force of the jet that keeps the ball up. Try a balloon on the air jet!

Theory: the rapidly moving stream of air forms a lower pressure region which keeps the balls within it.

Apparatus required: •Air jet •Retort stand and clamp •Various polystyrene balls •Table tennis ball •Light rubber ball •Beach ball.

5 Viscosity and falling dust

Relate the viscosity of the air to the speed of fall of light and heavy rain, builders' dust falling slowly in a room and the fallout after the Chernobyl disaster across Europe. Heavy rain hurts more than light rain — the droplets are not only bigger, they fall faster ($6\pi\eta rv = mg = \frac{4}{3}\pi r^3 \rho g$). Calculate the rate of fall of the different particle sizes. Try it with styrocell beads in water. Basically we are looking for the terminal velocity of the particles.

6   Styrocell beads in a snowstorm

A further demonstration of viscous drag can be made by pouring some styrocell beads into a jar full of water, putting on the lid and then turning the jar upside down. The beads fall slowly due to the viscosity of the water (Stokes' law) and simulate a fairly good snowstorm (Christmas fun!)

Apparatus required: •Styrocell beads •Jar with screw top lid (the taller the better; a plastic bottle will do) •Water.

7   Throwing a ruler (Bernoulli)

Throw a 30 cm ruler by holding it in the centre horizontally and then flipping it about its long axis. The resulting spinning motion should make it curve up or down. Compare this motion with top and bottom spin of a table tennis ball.

8   Blowing between two sheets of paper

One of the simplest demonstrations of reduced pressure due to fast moving air is simply to hold two sheets of paper vertically, one in each hand, and then to blow down through the gap between them. The lowering of the pressure in the moving air (Bernoulli effect) explains why they are drawn together.

Apparatus required: •Two sheets of paper.

9   Moving a ball in the air

The bending of the path in air of a spinning table tennis ball, tennis ball, cricket ball or rounders ball can be shown very clearly by the following experiment. Put a polystyrene ball or a table tennis ball into a cardboard tube with one end covered and with an internal diameter just a little greater than that of the ball itself.
    Holding the tube by one end, swing it smartly so that the ball is thrown out. The side of tube will give the ball a spin about the vertical axis and it should curve across the lab. Deflections of at least a metre in a distance of ten metres can easily be obtained. It is even better outside where you can throw it faster.

Theory: as the spinning ball moves through the air the flow of air past it will be more rapid on one side than on the other. This creates a difference

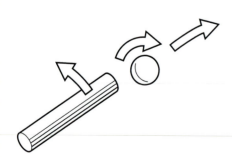

in pressure between the two sides of the ball and so the ball moves into the area of lower pressure.

Apparatus required: •Polystyrene ball or table tennis ball •Cardboard tube about 70 cm long.

10 Ball in funnel

Another simple demonstration of the lowering of pressure in fast moving air streams is to put a table tennis ball onto the bench and place a glass funnel over it. Attach the funnel to an air blower and switch on the air flow. The ball will be pushed up into the funnel and the rush of fast moving air between the ball and the funnel holds the ball in the funnel despite the fact that the funnel is held upside down!

Apparatus required: •Air blower •Table tennis ball •Glass funnel
•Rubber tubing.

SURFACE TENSION

General theory for this section

Pressure difference across a curved liquid–air surface of radius r is $2T/r$ where T is the surface tension of the liquid.

1　Wax on a kitchen sieve

This is a very good demonstration of surface tension. Immerse the sieve very briefly in hot wax to get a thin coating on the wires. Then pour water into the sieve — due to surface tension the water will not pass through the greasy holes. The angle of contact between the water and the wires has been increased to more than 90° and so the water will not wet the wires. Oil might work also (it will only work with really thick oil; cooking oil does not appear to be suitable).

Apparatus required: •Kitchen sieve •Wax •Suitable container to heat the wax.

2　Surface tension: glass block with water in between

Take two glass blocks and put a few drops of water on the surface of one of them and then press the two blocks together taking care to spread out the water film over the whole of the intervening surface. The very thin layer of water between the blocks gives a large pressure difference across the curved surface of the water between them and makes it very difficult to pull them apart.

Theory: pressure difference $= 2T/r$ where T is the surface tension of water and r is the radius of curvature of the water surface.

Apparatus required: •Two rectangular glass or acrylic blocks •Water.

3   Surface tension and soap solution

The effect of soap or methylated spirit on the surface tension of water can be shown very easily by this simple experiment. Cover the bottom of a tray with a thin layer of coloured water (1 mm or so). Touch the water surface with a glass rod dipped in spirit or soap solution. The water springs away leaving the bottom of the tray dry.

Apparatus required: •Tray •Water •Soap solution or methylated spirit.

4   Giant bubble recipe

A number of experiments concerning bubbles are more impressive if a 'strengthened' solution is used. To make these giant bubbles and sheets of soap film the following recipe is suggested. You can not only blow big bubbles (diameter over 30 cm) but also make wonderful soap films on a frame made from four plastic straws that can be made to oscillate, demonstrating slow SHM.

Apparatus required: •You will need to make a mixture of 100 ml of bubble bath, 400 ml of tap water (this may be a little too much), 200 ml of gelozone (available from health food shops) and 50 ml of glycerine.

5   Soap bubbles

Soap bubbles have a number of uses in school physics. They can be used in experiments on surface tension, in a simulation of Millikan's experiment, for the estimate of the size of a molecule as well as for interference effects. The beautiful colours in the films are due to the path differences for different colours of light.

An upper limit of the size of the soap molecule can be gained from the fact that soap films or soap bubbles go black when they are about to break; they are then so thin (less than one wavelength of light) that there is simply a phase shift of π at one face and no interference due to their thickness. See also the experiment with large bubbles in *Miscellaneous mechanics*, experiment 25.

6 Boat and surface tension

Take a rectangular plant tray and fill it with clean water. Find a light aluminium food tray that is almost as wide as the plant tray and float it on the water. Touch the water surface behind the boat with a finger moistened with a little soap solution. The boat should be pulled along the tank by the greater surface tension of the water in front of it.

Theory: Net accelerating force on the front of the boat $= LT$ where L is the length of the front of the food tray and T is the difference between the surface tension of the water and that of the soap.

Apparatus required: •Plant tray •Rectangular aluminium food tray •Water •Soap solution.

7 Rotating dish

A splendid soap film can be formed on a crystallizing dish which, as it is spun on a rotating table, will give different thicknesses of soap film (thinner near the centre and thicker towards the edges) and therefore produce different colours due to interference within the film. I would suggest using the improved giant bubble recipe (see experiment 5) for this and viewing the result with a TV camera if available.

Apparatus required: •Crystallizing dish •Rotating table and suitable power supply •Blu tac •Soap solution •TV camera (optional).

8 How full is a glass

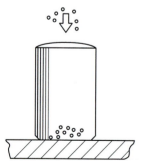

The effect of a convex meniscus can be shown by this simple experiment. Get one of the students to fill a glass with water. Now show that you can put in quite a bit more water, or some lead shot, without it overflowing; the convex meniscus keeps the water in place.

Apparatus required: •Glass •Water •Sand or lead shot.

9 Floating a needle

The ability of a water surface to support small objects such as insects can be demonstrated by using this experiment with a needle. Rest the needle on a piece of blotting paper on water and then lower them gently so that they float on the surface of water in a beaker. The filter paper eventually becomes water logged and then sinks leaving the needle floating. Adding soap will make the needle sink as it reduces the surface tension of the liquid. An alternative version is to try floating a razor blade.

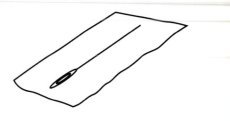

Theory: The water surface is depressed so that the needle is held up by the two surface tension forces ($2LT \cos A$ where L is the length of the needle and A is the angle that the water surface makes with the vertical where the needle rests on it). Weight of needle (mg) = $2TL \cos A$.

Apparatus required: •Blotting paper (the more absorbent the better) •Needle •Razor blade •Beaker of water •Soap solution.

10 The double bubble

The pressure difference in different sized bubbles can be shown by the following experiment. Join a length of glass tubing to each side of the top of a T-piece junction using a short length of rubber tubing with a tube clip over it. Blow a different sized bubble on each of the two ends of the T-piece of glass tubing in turn, closing the clip to seal off the bubble. When both bubbles are blown, open the clips to join the two but sealing off the arrangement from the air with a third clip at the base of the T-piece. The big bubble gets bigger and so less curved while the small bubble gets smaller and so more sharply curved — the pressure equalizes as the curvature of the two spherical surfaces becomes the same.

Theory: pressure difference across the curved surface of a soap bubble = $4T/r$ where T is the surface tension of the bubble liquid and r is the radius of the bubble.

Apparatus required: •T-piece •Glass tubing •Rubber tubing •Three tube clips •Soap solution.

11 Camphor boat

This experiment shows the lowering of the surface tension of water by a piece of camphor (a drop of soap solution will work equally well in its place). A small piece of camphor is put in a notch at the back of a piece of thick card shaped as a boat and floating on water. As the camphor dissolves the greater surface tension of the water at the front of the boat pulls it along.

Apparatus required: •Camphor granules or soap solution •Cardboard •Water in a sink or tank.

12 Diameter of a molecule

For the traditional oil drop experiment, you need a shallow tray filled with clean water; the cleanliness of the surface may be achieved by pulling a pair of waxed rods apart over the surface starting at the centre. A light dusting of lycopodium power is now sprinkled on the surface. A drop of oil is formed on a wire and its diameter ($2r$) measured. A TV camera used here instead of a travelling microscope makes life a lot easier. This drop is now placed on the water surface — it spreads out and the diameter ($2R$) of the resulting disc of oil formed on the water is measured. The thickness of the film (h) can then be calculated and the diameter of an oil molecule cannot be greater than this.

Theory: volume of oil film ($\pi R^2 h$) = volume of oil drop ($\frac{4}{3}(\pi r^3)$). Therefore $h = \frac{4}{3}(r^3/R^2)$.

Apparatus required: •Tray •Water •Powder •Oil •Wire holder •TV camera or travelling microscope or half mm scale •Waxed rods •Ruler.

MISCELLANEOUS MECHANICS

1. An interesting double pulley
2. Strength
3. Energy in a balloon
4. Sycamore seed propellor
5. Unrolling a carpet: whiplash
6. Viscosity of air with an oscillating table tennis ball
7. Satellite orbits
8. Planetary orbit
9. Straw and potato
10. Racing car tyres: different kinds of rubber
11. Lung volume
12. Rollers
13. Monkey and bananas
14. The torsion balance
15. Toy divers
16. Propeller on a wooden stick
17. Horseshoe and cocktail stick
18. A rubber sheet for gravitational fields
19. Air brakes — trolley with card sail
20. Simple pulley model
21. Centre of percussion of an object
22. Efficiency of a bicycle
23. Building bridges competitions
24. Perpetual motion
25. Rotating soap film again
26. Hunter on ice
27. Oil and vinegar
28. Picking up objects with your back to the wall
29. Rolling spool
30. Keys and matchbox over a dowel
31. Air brakes and a propellor
32. Large balloon
33. Three coins
34. Conservation of angular momentum

MECHANICS

1 An interesting double pulley

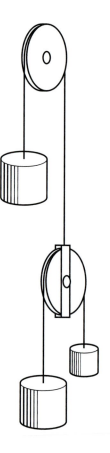

This intriguing problem uses two single pulleys. The lower pulley has a thread hung over it with a 100 g mass on one end and a 200 g mass on the other. This pulley is attached via a second thread over a second and upper pulley, which is fixed to a retort stand, to a mass of 300 g (with a little plasticine added to compensate for the mass of the first pulley). Since the masses at the two ends of the thread over the upper pulley are equal, ask the students what happens when the masses are allowed to move.

With the masses quoted here the large mass moves downwards with an acceleration of $g/17$.

To demonstrate this it is better to start off with the small masses being almost equal, otherwise their acceleration is too great.

Theory: Let the acceleration of the lower masses be $+a$ and $-a$ and the acceleration of the large (300 g) mass be A. This gives the acceleration of the free pulley as follows (for simplicity we will take g as $+10$ m s^{-2}):

Consider the 300 g mass:
$3-2T=0.3A$ assuming that it moves down — and it does!   (i)
Consider the 200 g mass:
$2-T=0.2(a-A)=0.2a-0.2A$ (ii)
Consider the 100 g mass:
$T-1=0.1(a+A)=0.1a+0.1A$ (iii)

Doubling (iii)
$2T-2=0.2a+0.2A$ (iv)

Now subtract (ii) from (iv)
$3T-4=0.4A$ (v)

Now multiply (i) by 3/2
$9/2-3T=[0.9/2)A]$ (vi)

Now add (v) to (vi)
$0.5=0.85A$

So $A=0.5/0.85=0.588$ m s^{-2}

Remember that we have taken g as 10 m s^{-2}, $10/17=0.588$, so the acceleration is $g/17$.

If you take g as -9.81 ms^{-2} then the acceleration becomes 0.577 or 0.58 m s^{-2}.

Apparatus required: •Two single pulleys •Set of slotted masses with three hangers •Thread •Retort stand •Clamp •TV and video to record the motion for later analysis or three light gates

2   Strength

Squeeze a set of bathroom scales to find the strength of your hands. Don't lean on it.

Apparatus required: •Bathroom scales

3   Energy in a balloon

Work out the energy in a blown-up balloon either by calculation or by the energy released and converted to kinetic energy on a linear air track, or by how much it takes to pump it up. If you pop it how much energy goes into kinetic energy of the fragments and how much into the sound?

What is the energy in a stretched rubber sheet? How much energy is stored in a gas when its volume and pressure change?

Apparatus required: •Balloons •Linear air track etc.

4   Sycamore seed propellor

This is simply a plastic propellor on a short stick. It demonstrates the principle of a propellor very well. Spinning it the right way by moving it between the palms of your hands makes it fly across the lab.

5   Unrolling a carpet: whiplash

A roll of carpet can be made to give a whiplash effect as you unroll it. It makes a useful introduction to the effects of whiplash of a head in car crashes, and hence the need for head rests in cars.

6   Viscosity of air with an oscillating table tennis ball

The viscosity of air and its damping nature can be investigated by using a table tennis ball fixed to a length of thread and used as a pendulum.

Apparatus required: •Table tennis ball •Metre rule •Retort stand and clamp •Thread.

7   Satellite orbits

The idea of this experiment is to draw a scale diagram for a satellite in orbit 200 km above the Earth's surface (with 1 mm representing 2 km), that is, with an orbit radius of 6600 km, or 3.3 m when scaled down. A section of orbit is drawn using a pencil fixed to a 3.3 m length of string and the tangent to the arc is drawn at one point.

The distance that the satellite would fall in 120 s is found using $s = 1/2gt^2$ (using $t = 120$ s and assuming that g is constant up to 200 km above the Earth's surface) and this is marked in the diagram. The distance round the orbit where such a drop occurs is measured and then the fraction of a whole orbit is worked out. The time for one whole orbit can then be worked out. It is important for the students to realize that the only piece of data that turns this into a real value for the satellite orbit from just a large drawing is the value for g.

Apparatus required: •Pencil •3.5 m of string •Large sheet of paper such as drawer lining paper •Metre ruler.

8 Planetary orbit

Use a conical pendulum to simulate a planetary orbit or, better still, the decay of a satellite orbit when it encounters the friction of the upper atmosphere.

The decay of an orbit can also be demonstrated by using a large glass funnel. The satellite is represented by a marble spiralling in — analogous to the action of a satellite being affected by air friction in the atmosphere.

9 Straw and potato

Get a potato and cut it in half. Then try and push a plastic straw through one of the halves from the flat side; it's no good — the straw just bends. Now drive the straw sharply into the flat side of the potato — it should now go straight through. A shout as you drive the straw through usually wakes up a sleepy class!

Apparatus required: •Potato •Plastic straw.

10 Racing car tyres: different kinds of rubber

Why is it that racing car tyres wear out so quickly? They feel sticky to the touch. In dry weather racing cars use slick tyres; these are smooth tyres with virtually no tread. At the start of the race the wheels are spun round rapidly, the tyres skid on the track, heat up and melt at the surface — this melted rubber gives a really good grip! They are useless in wet weather, however, when there is much reduced friction and therefore little heating.

11 Lung volume

It is possible to get an estimate of your lung volume by blowing into water filled measuring cylinders that have been inverted over a tank of water. The amount of water blown out gives an approximate value for your lung volume.

Apparatus required: •Sink full of water •A number of large measuring cylinders or cans.

---

12 Rollers

This experiment involving rolling two cylinders down a plane, one with a heavy axle and one with a heavy rim (both the same total mass) can be used as either an interesting demonstration for the younger students to make them think, or as an analysis of moments of inertia for the older students. The can with the heavy centre accelerates faster than the one with the heavy rim. Use tins loaded with extra mass at different points.

Theory: moment of inertia of a solid cylinder (the heavy axle) $= Mr^2/2$ where r is the radius of the axle.
Moment of inertia of the heavy rim $= Mr^2$ where r is the radius of the can.

Apparatus required: •Ramp •Two cans loaded at the centre or at the rim.

---

13 Monkey and bananas

A monkey hangs onto a weightless rope that is passing over a frictionless pulley. On the other end of the rope is a bunch of bananas of exactly the same mass as the monkey. What happens if the monkey begins to climb the rope towards the bananas?
   The bananas also move upwards with the same acceleration as the monkey. If the monkey now lets go of the rope both the bananas and the monkey fall, the distance between them remaining the same. If the monkey now grabs hold of the rope again — it may burn its hands! — they both come to rest.
   I am grateful to a colleague who suggested a further twist to this problem. What happens if the monkey reaches across and starts to eat the bananas?

---

14 The torsion balance

A home-made torsion balance made by a wire stretched between two supports with a light rod such as a plastic straw fixed to it at right angles can be used as an accurate balance. (A clamp can be a useful way of tensioning the wire.) It can also be used to demonstrate an accurate method of force measurement or investigation of the shear modulus of a metal.

Apparatus required: •Wire •Two retort stands bosses and clamps •G-clamps •Light rod or plastic straw •Slotted masses (light).

15   Toy divers

Some toyshops sell plastic toy divers which you can blow into via a tube to make them rise, or suck to make them sink. They are rather like variable cartesian divers. See the density, upthrust and Archimedes section (experiment 3).

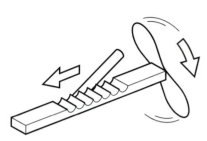

16   Propeller on a wooden stick

This wonderfully simple toy is just a serrated edged stick of about 1cm² cross section, about 20 cm long and with a thin wooden propeller nailed to one end but the physics in it is fairly complex. Rub a small piece of dowel along the serrated part and the propeller will rotate. Rubbing it along the other side can make the propeller rotate the other way.

Theory: elliptical standing waves are set up in the rod caused by the asymmetric vibrations due to the serrated edge.

17   Horseshoe and cocktail stick

A cardboard horseshoe stands resting on a cocktail stick as shown in the diagram. The problem is to pick up the horseshoe and the cocktail stick with just the other cocktail stick without touching anything with your hands except the second stick
    One possible solution: allow the horseshoe to fall forwards slightly so that the supporting stick protrudes below it, resting on the other stick. Then use this to pick up both the first cocktail stick and the horseshoe.

18   A rubber sheet for gravitational fields

A rubber sheet fixed over an old bicycle wheel (with the spokes removed) makes a very good demonstration of a simulated gravitational field. The sheet should be fixed to the wheel by a piece of string tied tightly around the rim. A set of slotted weights hanging from the centre of the sheet allows you to simulate the change in mass of the central object (planet or star) very easily. Then just roll in a ball bearing (to represent an orbiting satellite or planet), try to get it to move along a tangent to a circle about the central mass. Using a single heavy ball bearing in the centre rather than the slotted masses shows an effect on both the central mass and that approaching from the outside since the central mass moves.
    The depression of the sheet as a heavy ball bearing rolls across it simulates the curvature of space formed by the gravitational field of a massive object.
    An alternative method is to use a large coffee tin with the bottom cut

and with one end covered with a sheet of clingfilm. The whole apparatus can then be placed on an OHP so that is can be seen by the whole class.

Apparatus required: •Bicycle wheel with the spokes removed •Thin rubber sheet •Set of slotted masses •Various ball bearings

19 Air brakes: trolley with card sail

As an extension to the investigation of a toy car running down a ramp try using air brakes. Tie a piece of thread to the car and wrap the other end round a glass rod that stands in a test tube. Fix a piece of card (postcards are ideal) to the top of the rod to act as an air brake. As the car runs down the ramp the rod rotates thus rotating the card. Investigate the effect of different sized pieces of card on the acceleration of the car. Suggest plotting a graph of the square of the velocity after a given distance against the inverse of the area of the card.

Apparatus required: •Toy car •Cardboard •Timing mechanisms •Wooden board to act as a slope.

20 Simple pulley model

An interesting and surprising example of the mechanical advantage of a pulley system can be shown by the following demonstration. Two rods are held parallel by two people (a couple of retort stand rods will do fine for this). A piece of string is tied to one and then looped a few times round both. If you hold the other end of the string you will be able to pull the rods together no matter how hard the people holding the rods try to prevent you. If there is little friction between the rods and the string the more loops you can make the better it will work. You should see that the result is exactly analogous to a real pulley system with a large velocity ratio and hence a large mechanical advantage for a given efficiency.

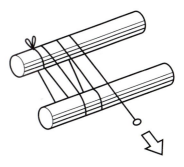

Theory: For a pulley system, efficiency = mechanical advantage/velocity ratio.

Apparatus required: •Two strong dowel rods •Pieces of broom handle or retort stand rods •Smooth string.

21 Centre of percussion of an object

Suspend a 1 m ruler from the 5 cm mark on a matchstick. Hit it sharply with a hammer two thirds of the way down the ruler from the match. What happens? Try this at a variety of points and relate to the forces felt by games players in cricket, baseball and rounders.

22 Efficiency of a bicycle

Mount the bicycle upside down and by measuring the mechanical advantage and its velocity ratio work out the efficiency of the bicycle. The mechanical advantage is found by hanging a weight on one of the pedals while finding out the force applied at the rim of the wheel to counteract this using a newton meter. The velocity ratio is calculated by measuring the distance moved by a point on the rim of the wheel for one rotation of the pedal.

Theory: efficiency of the bicycle = [mechanical advantage/velocity ratio $\times$ 100].

Apparatus required: •Bicycle •Sets of large slotted masses •Newton meter •String •Retort stands and clamps.

23 Building bridges competitions

This is worth doing both for general interest and for a study of the structures sections of syllabuses. It is a useful way of getting the students to think about moments and vector analysis. Plastic straws are quite good. Fixing them together is a problem — we have even sewn them together with thread!

Get them either to try to bridge the largest gap with the apparatus provided or make the strongest bridge to cross a gap of a certain width.

Apparatus required: •Plastic straws •Card •Adhesive tape •Thread

24 Perpetual motion

A disc of wood is mounted so that its centre is held by an axle in the wall of a tank of water, half the disc being in air and the other half in water. The side in the water experiences an upthrust (we will assume that there is a perfect frictionless seal so that no water leaks out). Why doesn't the disc rotate?

25 Rotating soap film again

Use a tin can with a soap film across the end and with its axis horizontal. Spin the can on a motor and illuminate it with a 100 W lamp from behind a translucent screen. As the can rotates the soap film does two things. It distorts, the shape being that of a rotating liquid surface, and it thins at the centre. It can therefore be used not only to demonstrate the liquid surface shape but also to look at the interference within a thin film of varying thickness. (See also surface tension experiment 7).

Apparatus required: •Tin can •Soap solution •Motor •100 W lamp •Translucent screen

26 Hunter on ice

This is an interesting example of equal and opposite forces. Imagine that you are standing on a frozen lake holding a rope, the other end of which is tied to a sleepy polar bear! You pull on the rope and if there is negligible friction you and the bear both move. As you would expect, because of its greater mass the bear accelerates less rapidly than you do but where do you meet? In fact it can be proved that if there is no friction you and the bear will always meet at the same place where $md_1 = Md_2$. Your mass is m and you move d_1 and the bear, mass M, moves d_2.

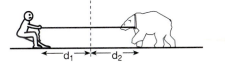

27 Oil and vinegar

An interesting example of flotation is the problem of different mixes of salad dressing poured from one bottle. A man and his wife go on a picnic but for their salad dressing they only take one bottle containing oil and vinegar; the oil floats on the vinegar. The problem is that they both like a different mix of oil and vinegar. How can they get this from just one bottle?

  Solution: pour off some oil and then turn the bottle upside down; the oil still floats on the vinegar, so if you put your thumb over the neck of the bottle as you turn it upside down, removing the thumb can let the required amount of vinegar out!

Apparatus required: •Narrow-necked bottle •Oil •Vinegar

28 Picking up objects and standing up: problems with balance

(a) An interesting example of balancing and the centre of mass of the human body can be shown by getting students to stand with their heels against a wall and then asking them to pick up an object placed, say, half a metre in front of them. This does not appear to be possible.

  (b) Another example of balance is to get students to sit upright on a straight-backed chair with their feet flat on the floor and the lower part of their legs vertical. Then ask them to stand up without leaning forward. It is impossible — they cannot do it unless they get some part of their body weight in front of the pivot point on the chair.

29 Rolling spool

This interesting and intriguing demonstration uses a spool of thread. A length of thread is unwound and is pulled gently with the thread at an angle to the horizontal. With the thread at a small angle to the ground the spool is pulled forwards, while if the angle is large it goes backwards. The critical point is when the line of the thread meets the point where the spool touches the bench.

30 Keys and matchbox over a dowel

Tie a bunch of keys to one end of a length of thread and a matchbox to the other. Loop the thread over a piece of dowel rod fixed horizontally in a boss. Hold the matchbox so that it is on the same level as the dowel rod with the keys hanging vertically downwards. Let go. The keys fall and the matchbox swings down but the thread wraps itself around the dowel rod.

Apparatus required: •Retort stand and boss •Wooden dowel rod •Matchbox •Bunch of keys or 50 g mass •Thread.

31 Air brakes and a propellor

A very good demonstration of air brakes is to make a windmill from a set of card vanes fixed in a cork and then mounted over a glass tube with a closed end on a glass rod. A thread wrapped round this is hung over a pulley and fixed to an accelerating mass or to a car on a ramp. Differing damping effects can be produced by changing the size of the vanes.

Apparatus required: •Cork •Cardboard •Glass tube •Glass rod and pivot •Thread •Car •Ramp •Pulley

32 Large balloon

To measure the density of air it is useful to use a large rubber balloon — one that you can blow up to a diameter of over 30 cm. The volume is about 0.03 m^3, therefore the mass of air is 36 g. Is it? The pressure of the air changes as the balloon is inflated and therefore the actual mass of air that is measured when it is weighed on a top pan balance is greater than the mass of the same volume at normal atmospheric pressure. It makes a good introduction for the younger students but can also be used by older students to study the gas laws.

Apparatus required: •Large balloon.

33 Three coins

The conservation of momentum and Newton's cradle can also be demonstrated by three coins on the table. Place two together, holding one down with your finger. Slide the third coin towards this coin so that a collision takes place. The third coin comes to rest while the first coin flies off.

A larger-scale version of this can be demonstrated in a game of croquet. Place two croquet balls in contact with each other and put your foot on top of one of them. Then strike this ball with the mallet — the free ball moves off.

34 Conservation of angular momentum

This can be demonstrated by a number of simple experiments.

(a) Stand someone on a rotatable table with their arms outstretched. Give them a gentle push so that they start rotating and then ask them to pull their arms into their sides; the rate of spin increases since their moment of inertia has decreased and the spin rate must get larger to conserve angular momentum. (It works even better if they hold a kilogram mass in each hand!)

(b) Use a conical pendulum fixed to the top of a broomstick. The pendulum bob is given a push so that it begins to move in a circle, the string wrapping itself round the stick and the rotation rate increasing as the radius of the orbit becomes less.

(c) A cylinder is fixed to a string which is suspended from a beam. The cylinder is held on its side and spun round. As it falls with its axis vertical the rate of spin increases.

(d) Stand a pupil on the rotating table and get them to throw a 1 kg mass to you.

(e) Place a battery-powered fan on an overhead projector. When the fan is switched on the blades rotate one way while the body of the fan rotates the other.

Apparatus required: •Rotatable table •Two kilogram masses
•Broomstick •Overhead projector •Fan •String and cylinder.

RESONANCE AND DAMPING

General theory for this section

Resonance. Any object, such a child's swing can be made to vibrate if it is given a push, but you find that if the frequency of the push is one particular value the amplitude of the oscillation of the swing builds up — this is known as resonance. Put another way — if the driving frequency is equal to the natural frequency of the oscillating system then resonance results.

Damping. There are two types of damping:
(a) internal — where the amplitude of oscillation of an object reduces due to internal forces;
(b) external — where the amplitude of oscillation of an object reduces due to external forces such as air or liquid.

Either of these may produce light damping where the oscillations die away very slowly, critical damping where they die away fairly quickly in less than one oscillation, or heavy damping where the object takes a long time to reach its equilibrium position without oscillating

1. Mechanical resonance: hacksaw blade
2. Coupled pendulums: Barton
3. Resonance between buildings
4. Standing waves
5. Resonance again
6. Resonance and earthquakes
7. Resonance in a rod
8. Coupled pendulums: two SHM
9. Air damping
10. Tacoma Narrows bridge
11. Resonance curves and damping
12. Feedback
13. Resonance: tambourines
14. Coupled oscillations and resonance: the Wilberforce pendulum

1 Mechanical resonance: hacksaw blade

Mechanical resonance can be demonstrated very well by using a 30 cm long hacksaw blade mounted vertically in a clamp. Oscillate it by using a vibration generator pressing against the lower part of the blade and load the top of the blade with lumps of modelling clay. To get the different resonance frequencies alter the frequency of the vibration generator until the hacksaw blade makes large oscillations. We use two 5 kg masses placed behind both the vibration generator and the clamp to keep the arrangement steady. This could also be done by clamping it all to the bench. Resonant frequencies in the 10–20 Hz range are common.

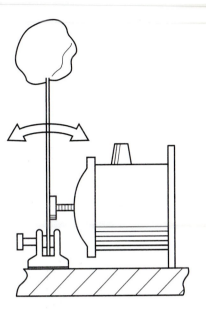

Another variation of this resonance experiment is to mount the hacksaw blade horizontally directly onto the vibration generator. It then becomes rather like the ruler twanged over the edge of the bench.

Apparatus required: •Vibration generator •Signal generator •Hacksaw blade •Two 5 kg masses or G-clamps •Two base clamps •Modelling clay.

2 Coupled pendulums: Barton

Set up a series of light pendulums all suspended from the same tight string. Fix one heavy pendulum to the string, the same length as one of the light pendulums. Now displace the heavy pendulum. Resonance will occur between the heavy pendulum and the light pendulum of the same length. All the pendulums will move with phase differences between them and the heavy pendulum. The traditional pendulums are lead spheres and paper cones but I have found that polystyrene balls are very suitable for the light pendulums.

Apparatus required: •Five light pendulums •One heavy pendulum •String •Two retort stands •Two G-clamps.

3 Resonance between buildings

I used to live in a house that was joined to my neighbour's by a garage. Each house had identical large open-plan lounges. When he used his stereo in his lounge I could hear it loudly in mine. The sound was transmitted along the joining beams and resonated in my house. The transmission of the bass frequencies was particularly unpleasant, the low frequency booming resonating in the large volume of the rooms!

4 Standing waves

As a demonstration of standing waves and resonance use a vibration generator mounted upside down connected to a signal generator. Hang a spring from it and attach a mass to the lower end of the spring. This

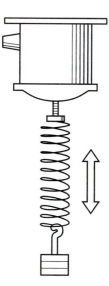

arrangement enables the resonant frequency of the system to be altered easily by changing the suspended mass. The resonant frequency can be found by changing the frequency of the vibration generator and looking for the largest oscillation of the mass. The period can be compared with that given by the formula for the natural period of oscillation of a spring, $T = 2\pi(e/g)^{1/2}$, where e is the extension of the spring when a mass m is hung on it at rest.

An alternative version of this experiment is to have the weight hanging from the spring which is then attached to a thread over a pulley coming up from the vibration generator

Apparatus required: •Pulley •Vibration generator •Signal generator •Thread •Helical spring •Slotted masses and hanger •Retort stand and clamps.

5 Resonance again

A further example of resonance using the vibration generator is to suspend a spring from a retort stand. Hang a weight on it and fix another spring below the weight. Fix this spring to a vibration generator and switch on. Adjust the frequency of the vibration generator to give resonance.

Apparatus required: •Vibration generator •Signal generator •50 g mass •Two short helical springs (3 cm × 1 cm unextended) •Retort stand •G-clamps.

6 Resonance and earthquakes

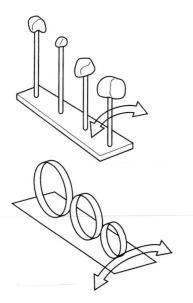

This experiment suggests two different methods of simulating the effect of an earthquake.

(a) A set of dowel rods of different lengths (or hacksaw blades) topped with different amounts of modelling clay is mounted in a board. The board can then be shaken to simulate an earthquake. Some of the rods will vibrate strongly showing resonance depending on the frequency of vibration of the board.

(b) The second method is to stick a set of paper rings of different diameters on a card. The rings represent different sizes of buildings. Then shake the card, if you get the shaking frequency right one ring resonates. Varying the shaking frequency will make different rings resonate. Rings of card and aluminium foil can also be tested to represent buildings made of different materials.

This makes a good simulation of the effect of earthquakes on different sizes, stiffness and types of buildings.

Apparatus required: •Dowel rods •Modelling clay •Board •Card •Paper •Aluminium foil.

7 Resonance in a rod

Take a metal rod, about 1 m long and with a diameter of 0.5 cm. Hold the rod in the centre with one hand and then stroke the rod with the thumb and forefinger of the other hand, having rubbed them with rosin. The rod will then 'sing'. To get a good resonance you will have to grip the rod quite tightly.

Apparatus required: •Metal rod 1m long and 0.5 cm in diameter •Rosin.

8 Coupled pendulums: two SHM

Suspend a filter funnel by four threads so that it is able to oscillate in two planes. Put your finger over the funnel outlet and fill it with sand. Now oscillate it over a sheet of paper. The pattern formed shows SHM in two directions.

Apparatus required: •Filter funnel •Thread •Sand •Retort stand and clamps.

9 Air damping

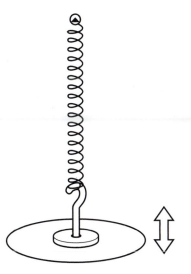

Suspend a mass from a spring. Fix a large cardboard disc above the mass. It is convenient to use the hanger from a set of slotted masses and have the disc resting in the base of the hanger with another mass on top to hold it in place. Pull the mass downwards and allow it to oscillate. You will need a spring with a spring constant that allows an initial amplitude of some 20 cm to show this effect well. As a demonstration, a slinky spring works very well. The effects of air damping can then be easily investigated by measuring the decay in amplitude as time passes. For the sixth form a graph of the natural log of the amplitude against the number of oscillations enables the equation of the motion to be found, namely $A = A_{o}e^{-kn}$

A further investigation can be carried out using discs of different diameters.

Apparatus required: •Helical spring •Slotted masses •Cardboard disc about 30 cm in diameter •Ruler in base clamp.

10 Tacoma Narrows bridge

This classic video clip of the collapse of this bridge is a must for all physics departments showing the resonance collapse of the bridge due to the effect of high winds on the special structure of the bridge. Note the man leaving his car. I understand that his pet dog was actually left in the car and fell with it into the river. Resonance can be a problem to all suspension bridges; in fact when I used to go to secondary school part of the route was across a small suspension footbridge over the river Severn in Shrewsbury. If two or three of us went across together a suitable heavy walking speed would get the bridge swinging. The local army cadets were told to fall out of step when crossing it. They didn't always obey orders!

11 Resonance curves and damping

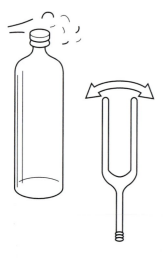

The nature of sound that can be made by blowing across the top of a milk bottle or twanging a tuning fork makes a good demonstration of the relation between the damping of an oscillating system and the shape of its resonance curve. The milk bottle has a broad resonance curve but heavy damping. The tuning forks have a sharp resonance but light damping and so the oscillations take a long while to decay. The decay of the sound can be measured with a microphone and a storage oscilloscope or with a computer sensor

Apparatus required: •Milk bottles •Tuning fork mounted on a resonant box •Microphone •Storage oscilloscope or computer sensor.

12 Feedback

Feedback is the phenomena where part or all of the output of a system is fed back to the input and affects it. A simple demonstration of this is as follows and is an effect that many of us have heard from the PA system at sports days or while trying to set up a sound system for a concert or play. Put a microphone in front of a speaker driven by an amplifier. The slightest noise will produce the high-pitched whistle denoting feedback.

There follows an analogy of positive feedback, where the feedback increases the output, and negative feedback, where the feedback decreases the output.

Positive feedback: a snowball rolling down a hill; the bigger it is, the bigger it gets. Negative feedback: a stream full of leaves or mud flowing through a narrow gap; the more leaves that get stuck the slower the water flows and the more leaves get stuck and so on.

13 Resonance: tambourines

Fix one tambourine by its rim in a clamp with its skin vertical and hang a light polystyrene ball so that the ball just touches the skin of the tambourine. Now hold a second identical tambourine parallel with the first and close to it. Strike this second tambourine. The resonance effect with the first should make the ball swing away from the skin.

Apparatus required: •Polystyrene ball on thread •Two tambourines •Retort stand and clamp.

14 Coupled oscillations and resonance: the Wilberforce pendulum

This device is simply a mass suspended by a spring. It has three modes of oscillation:
(a) a swinging motion,
(b) a vertical oscillation along the axis of the spring,
(c) a twisting motion.

If these last two modes can be made to have the same fundamental frequency, energy can be transferred between them. A way of varying the torsional oscillations without varying the mass is to have a wooden disc with four bolts imbedded in it and projecting radially. On each of these is a nut and if the nuts are moved in or out the moment of inertia of the disc is changed but not its mass.

Apparatus required: •Spring •Slotted masses •Wooden disc with four bolts and nuts fitted as described.

DOPPLER EFFECT

General theory for this section

This section deals with the Doppler effect – the change of observed frequency and wavelength produced by a moving source or observer. When a source of waves and an observer are moving toward or away from each other there is an observed change in frequency of the wave received by the observer. This change (Δf) is given by the equation $\Delta f = fv/c$ where v is the relative velocity of source and observer and c is the velocity of the waves.

1. Doppler velocity with microwaves
2. Doppler buzzer
3. The Doppler duck
4. Beats and the Doppler effect
5. A chocolate factory and the Doppler effect
6. Applications of the Doppler effect

1 Doppler velocity with microwaves

The frequency shift due to the reflection of a wave from a moving object (due to the Doppler effect) is very easy to do using the 2.8 cm microwave apparatus. Set up the transmitter and receiver side by side pointing the same way. The receiver should be connected to an amplifier and speaker. Put a sheet of cardboard in front of them mounted vertically in a base clamp. Now move a sheet of metal backwards and forwards on the other side of the card.

The Doppler shift is easily heard as a note from the loudspeaker. Swishing a metre ruler is especially effective. The faster you move the ruler the higher the pitch. It will work with just your hand and without the card if done in a lab with good reflections from the walls. The police radar speed trap frequency is apparently the same as that we all use in school with wavelength $\lambda = 2.8$ cm!

If the output is connected through the amplifier to an oscilloscope a very much more sensitive demonstration results; any movement in front of the card gives a large trace on the screen. The actual shape of the trace can then be seen easily.

Theory: the microwave transmitter produces a beam of microwaves, some of which is reflected from the stationary sheet of card into the receiver while some passes through and is reflected back to the receiver from the moving object. The waves from the moving object suffer a Doppler shift — an increase in frequency if the object is moving towards the card and a decrease if it is moving away. The two signals are fed to the amplifier and the difference signal is what appears at the speaker.

Frequency shift (Δf) = $2fv/c$

Apparatus required: •2.8 cm microwave transmitter •2.8 cm microwave receiver •Power supply •Loudspeaker •Amplifier unit •A4 cardboard sheet mounted vertically in holder •Metre ruler and metal plate •Oscilloscope •Optional TV camera and television.

2 Doppler buzzer

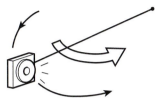

An easy way to show the Doppler effect with sound is to use a whirling piezoelectric buzzer. Connect a small buzzer to a battery or power supply by some long wires which should be taped securely to the buzzer, hold the battery in your hand and then swing the buzzer round your head. For wires a metre and a half long (or more) a really good Doppler effect is produced. You should ask the students whether they think that the person actually spinning the buzzer round will hear the Doppler effect or whether the change of pitch is just audible to somebody else.

It is also interesting to ask if it matters how close to someone you stand. Clearly it does: the nearer you are the more the interval between high and low pitch will be changed. (Instead of the piezoelectric buzzer you can use a small speaker.)

Theory: change of pitch $(\Delta f) = fv/c$. An emitted frequency of 300 Hz and a rotation rate of two per second in a one metre radius circle gives a frequency shift of 11.5 Hz, a change of pitch of a little under a semitone at that frequency.

Apparatus required: •Piezoelectric buzzer •Power supply or battery pack •Two 2 m leads •Electrical tape.

3 The Doppler duck

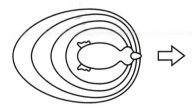

I use a clockwork duck with paddling feet to generate waves in a tank and to show the Doppler effect. As the duck moves along the ripples in front of it get compressed while those behind it are spread out — the longer the tank the better. I have often thought that a bath in the lab would be very useful. My duck keeps drowning! A TV camera mounted above the tank is especially helpful to show this effect to a large group.

Apparatus required: •Water tank •TV camera (optional) •Clockwork plastic duck.

4 Beats and the Doppler effect

The Doppler change of pitch can be used to give beats between two signals. For the two sources two signal generators are needed feeding outputs to two small speakers, one loudspeaker being mounted on a dynamics trolley (the rider on a linear air track is an alternative). If the frequencies of the two signal generators are adjusted to give beats then by moving the dynamics trolley the beat frequency can be made to change. This is an exact simulation of the police radar speed trap and is a sound version of experiment 1.

Apparatus required: •Two small loudspeakers •Dynamics trolley or linear air track •Two signal generators.

5 A chocolate factory and the Doppler effect

A good analogy to help to explain the Doppler effect is to imagine that you are working in a chocolate factory packing chocolates that come to you down a steadily moving conveyor belt. At the other end of the belt another person puts the chocolates on the belt at a steady rate. The chocolates therefore reach you at the same steady rate at which they were put on.

Now imagine that the other person starts to walk slowly towards you alongside the conveyor belt, still putting chocolates on at the original steady rate. You can see that you will receive the chocolates at a faster rate because after putting a chocolate on the belt your partner walks after it and when the next chocolate is put on the belt it will be closer to the first chocolate than if he or she had not moved. You will also receive chocolates at a faster rate if you walk towards the other end collecting chocolates as you go and your partner stays still.

Walking the other way will mean an increasing separation of chocolates and a drop in the rate at which you receive them – the Doppler red shift.

In this analogy the chocolates represent the crests of a wave, the rate at which they are put on the belt the original frequency of the source, the rate at which you receive them is the observed frequency, your speed (or your friend's speed) is the speed of the observer (or source) and the speed of the belt represents the speed of the waves.

6 Applications of the Doppler effect

1. Galactic red shift — lines in the spectra of receding galaxies shifted towards the red.
2. Plasma temperatures — broadening of spectral lines from fast moving atoms.
3. Doppler burglar alarm — reflected signal from a moving object.
4. Rotation of the Sun — shift in frequency due to recession and approach of solar limbs.
5. Radar speed trap — shift in frequency of the moving car.
6. Speed of blood flow in the body — shift in frequency of signal reflected from moving blood.

VIBRATIONS AND WAVES

General theory for this section

A vibration or oscillation is really just a wobble — a movement from side to side or up and down. Large, heavy and slack objects have a slow natural rate of vibration while small, light and tight objects vibrate rapidly.

1. Skipping rope: only certain frequencies
2. Wave motion
3. Waves and refraction
4. Standing waves on plates – the drum
5. Vibration
6. Standing waves
7. Standing waves on a vertical slinky
8. Melde with white elastic
9. Phase changes
10. Speed of waves along a rope

1 Skipping rope: only certain frequencies

Try swinging a skipping rope or wobbling a stretched rubber cord. There are only certain frequencies that are allowed. No matter how hard you try, all frequencies are not possible. The harmonics and standing waves on the rubber tube can easily be produced by using a circular rotation of the tube, rather like actual skipping (see the propeller on a wooden stick toy).

Theory: when the rope or cord vibrates with its fundamental frequency (the lowest) the length of the cord is one half of one wavelength. For higher modes of oscillation the length is equal to a whole number of half wavelengths.

Apparatus required: •Skipping rope •Rubber tubing.

2 Wave motion

The motion of a tuning fork and the sound waves produced by it can be observed using the following arrangement. Tape together three or four bar magnets all aligned the same way and put them in a coil of wire of around 3000 turns connected to a cathode ray oscilloscope.

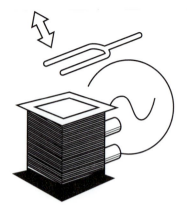

Now vibrate a tuning fork over the magnets. The fork becomes magnetized and so induces a potential in the coil which can be seen on the oscilloscope. Using a storage oscilloscope module will enable the waveform to be studied at leisure.

The beats formed between two tuning forks of similar frequency may also be studied using this method by striking them and then holding them both over the magnets.

Apparatus required: •Four strong bar magnets •Tape •Oscilloscope •3000-turn coil.

3 Waves and refraction

A useful way to explain the way light waves behave in reflection and refraction is to compare them to water waves. Water waves slow down as they pass from deep to shallow water and light waves slow down as they pass from air to glass. As waves move up a steadily sloping beach they slow down and their wavelength is reduced, the frequency of the waves remaining unaltered. If the angle of incidence is anything but zero (in other words if the waves hit the boundary at any other angle to it other than a right angle) then the waves will bend — refract — and their direction of travel will be changed. This is explained by realizing that one side of the wave front hits the join before the other and so slows down first so causing a change of direction. Driving a car from sand onto a tarmac road at an angle gives exactly the same turning effect.

This can be demonstrated by using a toy car running onto a line of sand on a perspex sheet. Total internal reflection can be seen!

Apparatus required: •Toy car •Perspex sheet •Sand •Overhead projector (optional).

4 Standing waves on plates: the drum

These can be studied using a metal plate fixed on the top of a vibration generator (even stiff card will do). Sprinkle some dust on the plate and then vary the frequency of oscillation of the vibration generator. Standing wave patterns should be seen on the plate. With the vibration generator switched off some interesting sounds may be heard when a musical box is placed on the plate and set going. The plate will resonate at certain frequencies.

Apparatus required: •Drum or stretched membrane •Vibration generator •Metal plate •Dust or talcum powder •Musical box

5 Vibration

The variation of pitch with vibrating mass can be clearly shown by the following simple experiment. Hold a ruler over the edge of the bench and twang it. The ruler vibrates, producing a note. The pitch of the note can be varied by sliding the ruler further on or off the bench. Changing from a wooden ruler to a plastic one demonstrates the dependence of stiffness of the vibrating object on the note produced. This can be used as a simple demonstration of vibration or for more advanced work to investigate the equation of a vibrating cantilever, using a metre ruler loaded with masses.

Plotting $\log T$ against either $\log M$ or $\log L$ where M is the mass on the ruler, L is the length of the ruler protruding over the bench and T is the period of oscillation will enable the student to predict the equation of the motion.

Apparatus required: •Rulers of different thicknesses and materials.

6 Standing waves

A small-scale version of standing waves in sound can be performed using a piezoelectric buzzer and a microscope slide. Set up the buzzer facing the slide and move the slide towards it and away from it, recording the node and antinode positions with a small microphone.

Apparatus required: •Piezoelectric buzzer •Microscope slide •Stands •Microphone.

7 Standing waves on a vertical slinky

Hang a slinky spring from a beam and fix the lower end to a vibration generator. By adjustment of the frequency of the vibration generator very good standing waves may be set up in the slinky. It is helpful to mark the nodes with coloured tape.

Theory: the distance between nodes (points of no vibration) is one half a wavelength.

Apparatus required: •Vibration generator •Signal generator •Slinky spring •Coloured sticky tape.

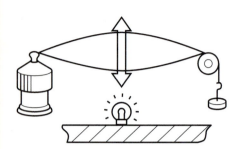

8 Melde with white elastic

The standing waves on a string are very well demonstrated by using a piece of white elastic fixed to a vibration generator, looped over a pulley and attached to a mass to give it tension. The vibration generator is switched on and the elastic is then viewed with a stroboscope. It is easy to change the tension by adding more mass.

(Health and safety: always ask the class about effects of flashing lights on them.)

Theory: Frequency $f = (^1/_2L)(T/m)^{1/2}$

Apparatus required: •Vibration generator •White elastic •Stroboscope •Slotted masses.

9 Phase changes

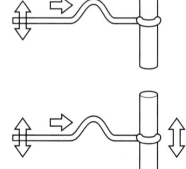

To demonstrate the phase change that occurs on reflection of a wave use two ropes. One with its far end fixed to a post and the other with a free end, such as a ring sliding along a vertical rod like a retort stand. A pulse is sent along each rope and is reflected from the far end. The phase change of π only occurs on reflection from the fixed end, a crest is reflected back as a trough. This is not the case with the free end nor is it for a water wave hitting the side of a harbour wall — the particles of water are free to move up and down and so no phase change occurs here.

Apparatus required: •Ropes •Broom handles.

10 Speed of waves along a rope

Mount two identical thick ropes side by side across the lab, one end of each rope fixed to the wall and the other hanging over a pulley with a mass on the end to tension the rope. Put a piece of folded card over each rope at the same distance from one end. Hit the ropes at the same point with a broom handle and observe the folded card. The rope with the greatest tension will carry the vibrations faster and the card on this rope will jump off first.

Theory: speed of waves along the rope $= (T/m)^{1/2}$ For a rope of mass per unit length $m = 0.1$ kg and tension 50 N the speed $= 22$ m s^{-1}, for tension $= 20$ N, speed $= 14$ m s^{-1}.

Apparatus required: •Two lengths of thick rope •Two pulleys •Masses •Card •Broom handle.

SOUND

General theory for this section

Sound is usually considered to be a vibration having a range of frequency that can be detected by the human ear. This is between about 50 Hz and 20 000 Hz but the top level of this range does decrease markedly with age. A difference can even be detected between the ages of 11 and 18.

1. Wine glass
2. Sea shell amplification
3. Wood blocks: musical scale
4. Playing a hosepipe
5. Playing a saw
6. Singing tube
7. Increase in pitch as a bottle is filled
8. Talking in helium
9. Chicken: resonance with rubbed string
10. Paper cup telephones
11. Velocity of sound: echo method
12. Difference tones
13. Reflection of sound
14. The refraction of sound
15. Bell in a bell jar
16. Sound transmission under water

1 Wine glass

Playing a wine glass makes an interesting introduction to sound and resonance. Partly fill a wine glass with water and moisten your finger with a little methylated spirit to get rid of any grease. Rub your finger round the rim of the glass to make it sing.

Good effects can be achieved if you apply a steady pressure between your finger and the glass, firmly enough to get the stick-slip motion needed to produce the sound but not enough to break the glass and cut yourself! Changing the amount of liquid in the glass gives a change in pitch of the note. A set of glasses with differing amounts of liquid can be used to play a tune.

Apparatus required: •Wine glass •Methylated spirit (or a little violin rosin).

2 Sea shell amplification

The sound of the sea is supposed to be able to be heard by holding a seashell up to your ears. Use a seashell, a large one if possible, and see if this is true. In actual fact what you are hearing is the effect of amplification of small sounds by the body of the shell. A tiny sound will be amplified by the shell, giving the rushing noise of the sea. Don't spoil the magic of the effect for the young by too much explanation!

Apparatus required: •Sea shell.

3 Wood blocks: musical scale

Cut up a series of hardwood blocks, with their dimensions as follows. All should be of the same cross section, 3 cm × 0.5 cm, and their lengths should be 22.0 cm, 22.8 cm, 24.2 cm, 25.8 cm, 27.2 cm, 28.3 cm, 29.5 cm and 30.5 cm. Drop them one at a time onto a hard floor so that one end hits the floor slightly before the other. The effect is like a xylophone.

4 Playing a hosepipe

The dependence of the length of a pipe on the fundamental pitch that it emits when played can be impressively demonstrated by this experiment. Fit a clarinet mouthpiece into one end of a hosepipe. Play an ascending scale by chopping off lengths of the pipe with a pair of loppers, having first calculated the length required for a good scale and marked them on the hosepipe. As it gets shorter the fundamental frequency of the pipe gets higher and so up goes the pitch. It has one big disadvantage as a musical instrument for the twenty first century — you can't go back down again unless you are very good at sticking hosepipe together!

Apparatus required: •Length of marked hose pipe •Clarinet mouthpiece •Loppers.

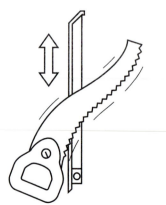

5 Playing a saw

A good example of the effect of tension on the notes produced by a musical instrument can be obtained by playing an ordinary saw. Hold the saw between your knees, bend it into an S shape and then play the smooth edge with a bow. Flexing the saw or your knees allows tunes to be played with luck and practice — I have managed part of Three Blind Mice!

It is worth emphasizing that there is no need for amplification – the large surface area of the saw means that a lot of air is set in motion and a loud and rather ethereal noise is produced.

Apparatus required: •Saw •Cello bow •Towel as protection for your legs.

6 Singing tube

There are two methods of producing sounds from a simple tube.

(a) One is just a plastic tube that you spin round your head and the sound you get is formed in the same way as blowing across the open end of a milk bottle. As the tube moves the rush of air across the open end gives a sound. Differing lengths resonate at differing frequencies.

(b) The other one is more complex. Set up a glass or metal tube (about 1 m long and with a diameter around 5 cm) vertically in a clamp. About 10 cm from the base fit a piece of copper gauze across the tube and then heat the gauze strongly until it glows red hot. Remove the flame. The tube should emit a loud singing note that lasts for some time. This is due to the expansion and contraction of the air in the tube as the copper cools down.

It is important not to allow the tube to heat up much, if it does so the effect is much less strong as all the air in the tube becomes hot and the contraction effects are not so pronounced. For this reason a metal tube is better than a glass one because it can be cooled quickly and the heat can be conducted away through its walls.

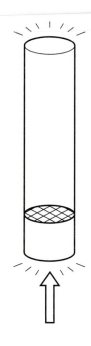

Theory: as the copper gauze cools the air around it contracts sharply and a shock wave is set up in the tube thus forcing the air within the tube into oscillation. Usually the fundamental note for the tube is produced ($L = \lambda/2$) but I have obtained harmonics ($L = \lambda$). The position of the copper gauze is fairly important, no effects being produced if it is too close to the end of the tube.

Apparatus required: •Plastic tube •Glass tube •Copper gauze •Bunsen •Retort stand and clamp.

7 Increase in pitch as a bottle is filled

The effect of the volume of a musical instrument on the pitch can be shown very easily by filling a bottle with water. As the water goes in the air in the bottle is excited and a sound is produced. This sound increases in pitch as the bottle gets fuller.

Apparatus required: •Milk bottle and a supply of water.

8 Talking in helium

If you breathe in small quantities of helium the effect of the lighter gas on the speed of sound in air, and therefore the pitch of the voice, can be studied. Always be very careful if you breathe in helium – you could suffocate without realizing it. On no account attempt to breathe in hydrogen — it is explosive!

The tendency of wind instruments to go sharp when taken into a warm concert hall is also due to this effect since the fundamental frequency of a pipe increases as the speed of sound in air increases and this happens when the temperature of the air rises. Notice that the density of the gas has no effect on the sounds heard from stringed instruments.

9 Chicken: resonance with rubbed string

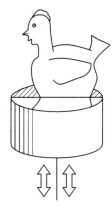

Take about 5 cm of cardboard tube (diameter also about 5 cm), and cover one end with stretched tracing paper. Fix a piece of thread through this. Make a chicken cut out that fits into the other end of the tube. Put rosin on your thumb and forefinger and, gripping fairly tightly, pull them down along the thread – a reasonable clucking and squawking sound can be made with practice! This is a good example of resonance and amplification.

Apparatus required: •Chicken made from cardboard tube and tracing paper •Rosin •Thread.

10 Paper cup telephones

Take two paper or plastic cups and make a hole in the base of each. Then thread a 3 m long piece of cotton through the hole to join the two together, fixing the cotton to the base of the cups with tape. Now pull the thread taught and speak into one of the cups, while someone else listens with their ear to the other one. The vibration of the air within the first cup is transmitted first to the cup and then to the thread. At the other end the vibration of the thread is passed to the second cup which then amplifies the sound.

My son and his girlfriend talked to each other between two buildings in a small village in Tuscany using this method. You can extend this to the idea of a loudspeaker by putting a large coffee tin on a vibrator to amplify the sound produced by a signal generator or radio.

Apparatus required: •Paper or plastic drinking cups or two small tins •Thread •Coffee tin •Vibration generator •Radio.

11 Velocity of sound: echo method

The measurement of the speed of sound is easy to do using the echo method. Fire a starting pistol outside in front of a distant building and record the time it takes for the echo to return. I use the school wall with a distance of nearly 200 m giving a 'there and back' distance of 400 m. The advantage of this method is that it uses not only a larger distance for the sound to travel, giving a longer time interval which is easier to measure but also eliminates wind errors. A minor additional detail: it also keeps all the class together in one place, both those doing the timing and the person firing the gun!

The discussion of the result should lead on to mentioning that the temperature of the air also affects the speed of sound; it is faster in hot air since the molecules are moving faster.

Theory: velocity of sound in a gas depends on the velocity of the molecules of that gas (v) since the sound waves are transmitted by the motion and collision of the gas molecules.

This is given by the equation $\frac{1}{2}(mv^2) = \frac{3}{2}kT$ so $v = (3kT/m)^{1/2}$ where k is the Boltzmann constant $(1.38\times10^{-23}\text{ J K}^{-1})$.

Therefore the speed of sound in a gas depends on (i) the absolute temperature T and (ii) the mass of the individual molecules of that gas, in other words the type of gas. Sound waves therefore travel faster through a hot gas which has light molecules.

Apparatus required: •Long measuring tape •Stop clocks •Starting pistol or clapper boards.

12 Difference tones

The superposition of sound can be shown by using a referee's whistle which has two holes, one on either side, or an old police whistle. (I have one from the 1940s given to me by my grandmother). Blow into it and you will hear a note that is the combination of the tones from the two sides. It gives a minor third output. Now bring your finger up to one of these holes to reduce the volume from that side and finally to cut it out altogether — you will hear a change in the overall pitch.

Theory: superposition of two simple harmonic motions $y_1 = a\sin(2\pi f_1 t)$ and $y_2 = a\sin(2\pi f_2 t)$
Final waveform: amplitude $(y) = 2a\cos2\pi(f_1-f_2)t/2\ \sin(2\pi(f_1+f_2)t/2$.

Apparatus required: •Two-hole referee's whistle or police whistle.

13 Reflection of sound

You can demonstrate this using two large concave mirrors with a small speaker at the focus of one and a microphone at the focus of the other. The intensity received by the microphone is then shown using an amplifier and an oscilloscope. Alternatively just put your ears at the focus of the second mirror.

Even simpler is to hold a large concave mirror up in front of you and move it inwards towards your face as you speak. When the distance between your mouth and the mirror is equal to the radius of curvature of the mirror a loud sound will be received by your ears. This is not strictly exact since your ears and mouth are not both at the centre of curvature of the mirror.

Apparatus required: •Two large concave mirrors, diameter around 45 cm •Microphone •Amplifier •Oscilloscope.

14   The refraction of sound

This can be clearly demonstrated using a signal generator, a balloon filled with carbon dioxide, a microphone and an oscilloscope. The balloon acts as a lens, focusing the sound.

Apparatus required: •Balloon •CO_2 supply •Signal generator •Microphone •Oscilloscope.

15   Bell in a bell jar

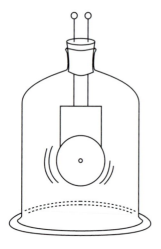

To demonstrate that sound needs a medium through which to travel, suspend a bell in a bell jar and then evacuate it using a vacuum pump. This has the disadvantage that the bell vibrates mechanically sending vibrations along the connecting wires. Standing the bell jar base plate on a sheet of foam rubber helps to reduce vibrations through the bench. As an alternative use a piezoelectric buzzer suspended by threads and connected to a battery within the bell jar.

   It would be good if the experiment could be done in a vessel that could be evacuated simply by allowing water to run out.

Apparatus required: •Bell •Bell jar •Vacuum pump •Sheet of foam rubber •Piezoelectric buzzer •Wire and battery •Thread.

16   Sound transmission under water

Lie in a bath and tap the side of the bath under the water with either your feet or a piece of wood. Now turn onto your side so that one of your ears is below the water surface. The sound will be much louder in that ear. The greater density of water means that the molecular vibrations are transmitted much more effectively through it.

Apparatus required: •Bath •Water.

GEOMETRICAL OPTICS

General theory for this section

Geometrical optics is the study of reflection and refraction of visible light. Light passing from one transparent medium to another (refraction) suffers a change in its velocity (violet light being slowed more than red) but there is no velocity change on reflection — just a loss of radiant energy.

1. Optics and the smoke box
2. Sources of light demonstration lesson
3. The camera obscura
4. Variable power lens
5. Distorted drawing
6. The silvered bicycle reflector
7. Metallized film: one-way mirrors
8. Pepper's ghost
9. Light reflection in a cube
10. Bent pencil in water
11. Refractive index: TV camera
12. Christmas tree balls
13. Image position
14. Bending of wood: measurement with a light beam; the optical lever
15. Non-lateral inversion: mirrors at 90°
16. Real/apparent depth: sideways block
17. Pinhole camera: 360°
18. A floating image
19. Interesting refractive indices

TOTAL INTERNAL REFLECTION

General theory for this section:

When light meets the boundary between two transparent materials it will be refracted but if it is travelling in the material with the greater refractive index and the angle of incidence is greater than the critical angle c it will be reflected back into the first material.

The critical angle is defined by the equation $n = 1/\sin c$ where n is the refractive index of the first medium (this simplified equation assumes the second material to be air).

20. Colour of tanned legs under water
21. Critical angle: semicircular block
22. Fibre optics

Geometrical optics

1 Optics and the smoke box

Shine light through a plate with holes in it into a glass-fronted box full of smoke. The path of the rays through various lenses placed in the box can easily be followed. Smoke can easily be made by a piece of burning sacking, rag or corrugated cardboard in a smoke generator (available from beekeepers).

Apparatus required: •Smoke-tight box with one glass side •Smoke generator •Large (10 cm diameter) lenses.

2 Sources of light demonstration lesson

I have found this an attractive way of introducing the topic of optics. We 'look' at a number of different sources of light such as the Sun, matches, a bunsen flame, iron filings and magnesium in the flame, a sodium chloride stick in the flame, a glowing wire, an electric light bulb, a low-energy discharge tube, gas discharge tubes (hydrogen, cadmium, sodium, neon, xenon, and helium), an LED, a laser and an ultraviolet light. The different colours of the sources, the time that they shine for and the energy losses are all considered.

Impressive effects using ultraviolet light can be obtained from shining it on silly putty, washing powder, nails, stage make-up, luminous watches and toys and security pens.

 Safety: take the obvious precautions with your eyes and those of the students. Wear protective goggles if needed and do not allow yourself or the students to look at the ultraviolet light or the laser directly or by reflection.

Apparatus required: •Laser •Matches •Bunsen •Iron filings •Magnesium •Electric light bulb •Gas discharge tubes •Low-energy lamp •LED •Ultraviolet light •Goggles.

3 The camera obscura

Small holes in shutters or blinds can give amazing views of outside scenes on the wall. It shows the pinhole camera on a very large scale. It is usually just the round disc of the Sun that can be seen but actual views can sometimes be obtained. I have seen one formed through a small hole in a blind in a hotel room in Sorrento in Italy. An inverted image of a whole mountainside was formed on my bedroom wall as the sun rose over the hills near Vesuvius. Images of the Sun can often be seen in a wood where the sunlight reaches the ground after passing through small holes between overlapping leaves.

Apparatus required: •Sunlight •Room with a fine hole in a blind.

4 Variable power lens

Use a plastic bag containing liquid, the curvature of which can be varied by squashing and stretching, to simulate the way the muscles of the eye change the shape of the eye lens. Experiments will suggest the correct sort of radius of curvature to give you a reasonable focal length.

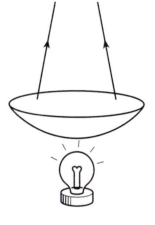

A very large version of this can be made by using a piece of transparent plastic or rubber sheet stretched tightly over the rim of a bicycle wheel. Pouring water into the top makes the plastic sag and if a lamp is put below the sheet an image can be formed on the ceiling of the lab. The sheet needs to be thin enough to distend with a relatively small amount of water otherwise the focal length is too great for any room but a high hall, like a sports hall.

Theory: the focal length f of a lens with spherical surfaces of radius r_1 and r_2 and of material of refractive index n placed in air is given by the equation: $1/f = (n-1)[1/r_1 + 1/r_2]$.

Apparatus required: •Plastic bag •Bicycle wheel without spokes •Thin rubber sheet (as clear as possible) •Lamp.

5 Distorted drawing

Two interesting phenomena can be observed in the following exercises.

(a) Make a drawing of an object by looking at its reflection in a curved surface, such as a polished tin can, and then view the drawing the same way.

(b) Make a distorted drawing so that it is only clear when viewed from an angle to the picture. A very good example of this is the skull in Holbein's painting, *The Ambassadors*.

6 The silvered bicycle reflector

Get a clear plastic bicycle reflector, made of pyramids of plastic and silver paint one side (spraying it with aluminium paint is fine). Look into it — you see a completely black reflector. This is simply the reflection of your retina. Since each 'hole' between the pyramids acts like a corner of a reflecting cube the light is simply reversed in direction. (I am grateful to Roy Sambles, professor of physics at Exeter University for this fascinating demonstration.)

7 Metallized film: one-way mirrors

This reflective film has a number of uses. The best source of it is in shops selling wrapping paper; they often have some in a variety of colours and I have even seen some with reflection diffraction gratings. In optics it makes a very good one-way mirror depending on which side is illuminated more strongly. When viewed from the highly illuminated side it looks like a mirror but when viewed from the less well illuminated side it is possible to see through it to the brightly lit area. It is also used by marathon runners to conserve heat after a race and so can be part of heat experiments.

Apparatus required: •Sheet of metallised plastic •Light dependent resistor (LDR) for light transmission experiments.

8 Pepper's ghost

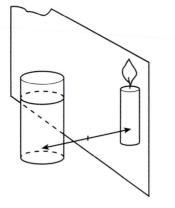

This is the reflection of a flame or a ghostly image from a sheet of glass. By this means a candle flame can be made to appear as if it is burning inside a glass of water. In fact it is simply that the image of the candle flame is at the same position as the water filled beaker. The cleaner and bigger the glass sheet available for this the better. A suitably edged piece of window glass is ideal. (Additional ideas appear in *School Science Review* March 1987, p 444.)

Apparatus required: •Beaker of water •candle •glass sheet.

9 Light reflection in a cube

Note how light behaves when reflected from the inside of the corner of a cube. It is reversed in direction irrespective of the original direction of the incident ray of light. Three plane mirrors can easily be fixed together to give a right-angled corner. This explains why reflectors on bicycles are made like this — the light from a car's headlamps is reflected back to the driver no matter from what angle it hits the reflector.

10 Bent pencil in water

One of the simplest methods of showing refraction in a liquid is to put a pencil in a glass of water and look at it from the top. The effect of refraction is clearly seen. The pencil appears to be bent – the end of the pencil seeming much nearer the surface than it should be. This can used to explain the difficulty of reaching into water to pick something out; your judgement of direction is impaired by the refraction.

Apparatus required: •Beaker of water •Pencil.

11 Refractive index: TV camera

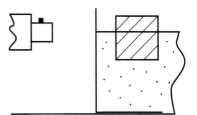

Float an object in water in a flat-sided plastic tank and view from outside with the TV camera placed so that it looks precisely along the water surface. Some of the block will be seen through the air and some through the water. The refractive index may be worked out by measuring the dimensions of the two views of the block from the TV screen.

Apparatus required: •Rectangular plastic tank •TV camera •Wooden block to float in the tank.

12 Christmas tree balls

It is well known that the formulae for curved mirrors only apply if the reflecting surface is restricted to the centre of the mirror, that is, to rays close to the axis. A good example of the distortion produced if this condition is not fulfilled, and when there is a reflection from a large part of a spherical surface, can be seen by using the coloured balls used for Christmas tree decorations. Using a TV camera would show the effects to a large audience.

Apparatus required: Christmas tree balls •TV camera if available

13 Image position

Mount a small plane mirror vertically on the bench with a board writer standing in front of it (the board writers used should be a centimetre or two higher than the mirror). You will see the image of the board writer in the mirror. Move a second board writer until the top of it seems to coincide with the image of the first one from whatever direction it is viewed (there is no parallax between them). The second board writer is now at the image position of the first.

Apparatus required: •Plane mirror mounted vertically •Two board writers or pencils.

14 Bending of wood: measurement with a light beam; the optical lever

Use an optical lever to detect the bending of a bench. This can be done by fixing a small mirror to the bench and then reflecting a beam of light from it so that a spot is formed on the wall. It works especially well if the scale is on the other side of the lab. Now lean on the bench. The beam of light is deflected through twice the angle of deflection of the table and the spot moves. This method can also be used to detect the bending of walls if someone leans on them or the deflection of some ceilings if somebody walks overhead. The use of a light beam as a pointer is especially good since it has 'zero' mass.

Apparatus required: •Plane mirror •Laser or fine beam of light •Ruler.

15 Non-lateral inversion: mirrors at 90°

If you look at the image across the join of two mirrors that have been placed at 90° to each other you will not see any lateral inversion.

Apparatus required: •Two plane mirrors fixed at 90° to each other •TV camera if available.

16 Real/apparent depth: sideways block

This experiment uses the real and apparent depth principle in a simple way to determine the refractive index of a block. Lie a glass or acrylic block on its side on a sheet of paper and look at the base of the block through one side. It is then easy to mark the apparent depth of the block by measuring the length of the shiny part on the sheet of paper under the block.

Theory: refractive index of the glass block = real depth/apparent depth.

Apparatus required: •Glass block •Sheet of paper •Pencil.

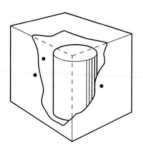

17 Pinhole camera: 360°

It is simple to take a photograph with a pinhole camera by substituting the paper screen by a piece of photographic paper. However, an interesting extension of this is to take an all-round photograph. Mount a toilet roll in the middle of a box with a pinhole in each of the four vertical faces of the box. In the dark room, or in a light proof bag, wrap photographic paper around the toilet roll tube.

Expose the film outdoors by uncovering the holes for up to 30 s in overcast conditions but around 10 s in sunlight. Develop and print the paper. A 360° picture should be obtained. A positive print can be formed by placing the negative print face downwards on the bench in the darkroom using only the safe light on top of another sheet of photographic paper with its sensitive side uppermost. Then switch on the ordinary light for about 30 seconds and print as normal.

Apparatus required: •Pinhole camera with centre mounted cylinder •Photographic paper •Darkroom and chemicals.

18 A floating image

Wave a white stick in mid-air at the focal plane of a slide projector. The image on the slide will appear hanging in space, demonstrating the persistence of vision of the human eye.

19 Interesting refractive indices

Some liquids have a refractive index almost equal to that of glass (about 1.5). Glass objects put in these liquids will seem to disappear. Glycerine, $n = 1.47$; castor oil, $n = 1.48$; xylene, $n = 1.51$.

Total internal reflection

20 Colour of tanned legs under water

Have you ever noticed why tanned legs look less tanned as they go from air to water? Is this a layer of air bubbles on them? Probably not — it is likely that it is just due to light entering the water and then being internally reflected; some of the light will be 'trapped' within the water, so giving the legs a brighter appearance there.

21 Critical angle: semicircular block

Shine the light from a ray box or a laser into the semicircular block directed at the centre of the flat side. It therefore enters at right angles to the curved side and so suffers no refraction at this point. Twist the block about the centre of the flat side so as to increase the angle of incidence of the light within the block with this side and so find the critical angle. Further increase will give total internal reflection.

Apparatus required: •Ray box •Semicircular perspex block.

22 Fibre optics

Some uses of fibre optics are: communications, endoscopy, security fences (you can't cut the optical fibre without the security guard knowing if they are watching a TV programme which has been transmitted down it!), lighting models.

23 Light along a water jet

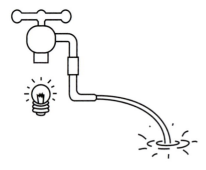

This is a pretty example of total internal reflection. Get a piece of glass tube bent into a right angle to use as a water jet. Mount it in a clamp and attach it to a water tap by a length of rubber tubing. Set up a small lamp just behind the bend and switch on both the flow of water and the lamp. The light will flow along the water path giving a gleaming patch where the water hits the base of the sink. It behaves just the same as the internal reflection along an optical fibre.

Apparatus required: •Light source •Water jet from L-shaped glass tube •Clamp •Sink or bucket.

24 Total internal reflection: corner of a glass block

Show that you cannot get light to cross the right-angled corner of a glass block. Looking across it will simply show you the apparently silvered surface at the other face.

25 Object and image

To show lateral inversion in a plane mirror use a pair of model cars, one British and one continental with their steering wheels on different sides. Reflect them in a plane mirror to show that they appear laterally inverted.

26 Total internal reflection and the TV camera

Use a rectangular straight-sided plastic tank with some water in it (the one used for wave demonstrations is ideal). Point the camera upwards towards the underside of the water surface from the outside of the tank. A splendid silvery surface can be seen.

Apparatus required: •Rectangular-sided plastic box •TV camera •Laboratory jack •TV.

27 Total internal reflection and soot

(a) This can be demonstrated by using a test tube immersed in water. The sides of the tube look silvery. A better and more controllable version is to use a glass tube. Put the tube in the water while holding your finger over the upper end to prevent the tube filling with water. It is full of air and so looks silvery just like the test tube, but when you take your finger off the water flows in and the silvery effect disappears.

(b) A further example of total internal reflection can be shown by blackening an egg with soot from a candle flame. Immerse the egg in water; it looks silvery because of the thin film of air that is trapped within the soot only to be seen to be black when it is lifted out.

(c) Another method is to use a table tennis ball and cover it with soot. When placed in water it will appear silvery due to the air film that clings to the soot giving total internal reflection.

Apparatus required: (a) •Beaker •Test tube •Glass tube (b) •Egg •Beaker of water •Candle.

28 Mirror in tank: refraction

The following experiment shows a good method of demonstrating both refraction and total internal reflection. You will need a rectangular tank with water in it to which some fluorescein has been added to make the light beams visible. Have a mirror held by a thread mounted under water in one end of a rectangular tank. A second mirror fixed above the tank enables you to direct a ray down onto the immersed mirror. The angle at which the ray of light hits the underneath of the water surface can easily be altered by pulling on the thread. Refraction at the air–water surface can be shown by using only the second mirror while both are needed for total internal reflection.

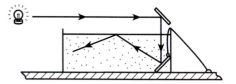

Apparatus required: •Rectangular plastic tank •Two plane mirrors •Fine beam of light from a high intensity light source or a laser •Thread •Fluorescein •Water •Modelling clay

29 Mirage

A good demonstration of a mirage can be obtained by burying a few (up to six) immersion heaters in a tray of sand. Switching these on will soon give a mirage in the hot air above the sand.

Apparatus required: •Six low voltage immersion heaters •Six power supplies.

30   The floating coin

Put a coin onto the base of a bowl and fix a cardboard tube at an angle so that it points above the coin and so that the coin cannot be seen through it (alternatively direct a TV camera to view over the coin).

Now add water to the bowl; the coin will mysteriously float up into view due to the refraction of light as it emerges from the water into the air.

Apparatus required: •Coin •Bowl •Cardboard tube and stand •TV camera (optional).

INTERFERENCE

General theory for this section
This section deals with the phenomena of interference with both sound and light.

Constructive interference occurs when the path difference between two wave trains is a whole number of wavelengths. A phase change of π occurs at a reflection with an optically more dense boundary, i.e. air to glass.

1. Interference with sound
2. Interference and distance measurement
3. Interference and butterflies' wings

1 Interference with sound

Interference with sound can be demonstrated by setting up two loud speakers connected to the same signal generator. Get the students to walk round the lab passing the speakers as they do so; a good interference pattern will be observed. They will notice that there are some places where the sound is loud and others where it is relatively soft. (A wavelength of about 0.75 m (frequency 440 Hz) is quite good.) Smaller wavelengths give good effects if you simply sit still and move your head from side to side. The experiment would be better outside, where the results would not be affected by reflection from the lab walls. An alternative is to detect the maxima and minima using a microphone which can be moved in front of the speakers on a long rod such as a metre rule. The output of the microphone can be fed through an amplifier to a CRO and the amplitude of the resulting trace measured.

As an extension of this, take two stereo speakers, reverse the phase of one and point them towards each other — the bass tends to cancel.

Theory: if the path difference between the speakers and the student's ear is a whole number of wavelengths the waves will add up and they will hear a maximum. If it is an odd number of half wavelengths destructive interference will take place and the sound will be much quieter.

Apparatus required: •Signal generator •Two loudspeakers •Microphone •Oscilloscope •Amplifier.

2   Interference and distance measurement

Use a pair of glass plates illuminated by sodium light to show the interference patterns due to the gap between the plates. Place the plates on top of each other on the bench and illuminate them from above with light from a sodium lamp. Irregular interference fringes can be seen due to the changing separation between the plates. Pressing on the top plate will alter these patterns, especially if the two plates are separated at one end by a piece of tissue paper. A tiny piece of grit between the plates will show circular interference rings (the TV camera is especially useful here).

Theory: the path difference between light reflected at the lower surface of the upper plate and the upper surface of the lower plate is $2dn$ where d is the separation of the plates and n is the refractive index of the air.

Apparatus required: •Sodium lamp •Two glass plates •Tissue paper.

3   Interference and butterflies' wings

The beautiful colours on butterflies' wings are due to the interference of light reflected from their surface. That these colours really are due to this effect can be shown by placing a few drops of a clear liquid such as acetone onto the wing surface. The liquid will fill up the gaps between the ridges on the scales and so change the colour. Viewing the wings with different colours of light also shows some interesting effects.

The Christmas tree structure in the morpho group of butterfly wings is especially spectacular, iridescent blue colours being formed. You might also try similar experiments using feathers — peacocks' tail feathers are particularly good.

Apparatus required: •Butterfly wing •Acetone and dropper •White light source •Filters.

DIFFRACTION

General theory for this section
This set of experiments deals with diffraction. Experiments using light, sound and microwaves are discussed.
Diffraction is the bending of radiation around obstacles or through apertures. Diffraction can also occur when waves reflect from an uneven surface. The greater the wavelength or the smaller the obstacle or aperture the greater the diffraction.

1. CD and DVD diffraction
2. Fingers and diffraction
3. Diffraction through fork prongs
4. Acoustic diffraction grating
5. Diffraction with sound
6. Diffraction and condensation
7. Resolving power
8. Diffraction with sound
9. Diffraction through tights

1 CD and DVD diffraction

(a) A CD will give a lovely diffraction pattern due to the tracks on it acting as a reflection diffraction grating. The same effects may be obtained from a peacock wing, dragonflys' wing and iridescent beetles. There is no colour at all — the peacock feathers split up white light to make the brilliant colours!

(b) If you use a digital video disc (DVD) the width of the diffraction pattern is increased. This occurs because the narrower tracks on the higher density DVD act as smaller obstacles and so give broader diffraction patterns.

(c) An extension of (b) is to use a laser with the CD placed on the bench. Direct the laser so that it hits the CD at grazing incidence. Take special care of beams that may be reflected in various directions.

Theory: $\lambda = d \sin\theta$; for a smaller grating spacing d the greater the value of the angle of diffraction θ for a given wavelength λ.

Apparatus required: •CD or DVD •White light source such as the laboratory lights •Gas discharge tube •Laser for experiment (c).

2 Fingers and diffraction

Diffraction can be observed using nothing more than your fingers! Put two fingers together upright in front of your eyes and look at a light source through the small gap between them. Fine dark diffraction lines can be seen due to the light spreading due to diffraction in the narrow slit. It works best using a straight fluorescent tube.

Apparatus required: •Light source.

3 Diffraction through fork prongs

This is another very simple demonstration of diffraction using the same principle as that described in experiment 2. Hold up a fork and look at a light through the prongs — dark diffraction lines can be seen in the spaces between the prongs. Rotate the fork parallel to the prongs to give an effectively smaller gap and so wider fringes.

Apparatus required: •Fork.

4 Acoustic diffraction grating

An acoustic diffraction grating will show diffraction of sound. One of these can be made using a large cardboard tube with a row of 1 cm diameter holes running from end to end. Put a small loudspeaker at one end and adjust to give a frequency of, say, 500 Hz. Now move a microphone along parallel to the tube and observe the rise and fall in sound intensity by looking at the microphone output on an oscilloscope.

Apparatus required: •Cardboard tube •Loudspeaker and amplifier •Microphone and oscilloscope.

5 Diffraction with sound

Diffraction with sound waves can also be shown easily by using a loudspeaker to produce a sound wave and then getting one of the students to align himself or herself behind another's head. The diffraction effects can then be observed by moving their own head from side to side. The sound waves spread round the obstacle (the student's head) and maxima and minima can be detected. Using a wavelength of about 10 cm (3 kHz) shows the diffraction round the head really well.

Apparatus required: •Pupils •Signal generator and loudspeaker.

6 Diffraction and condensation

Misted up glasses give good diffraction effects! I discovered this while trying to read a book in the bath with a cold pair of glasses. Viewing a light bulb through the fine mist of water vapour on the glasses showed coloured diffraction rings. The physics of condensation is also demonstrated here! Hot air can contain more moisture than cold air and so when someone puts on a pair of cold glasses in a hot bath the air near the glasses is cooled and some of the water vapour it contained condenses onto the glasses. I have also noticed the effect in my car, which was left outside on a cold night after being used the previous day with the heater on. The air in the car had been warm and so could contain a reasonable amount of water vapour but overnight it cooled and this water condensed out onto the inside of the windscreen.

Apparatus required: •Glasses •Kettle.

7 Resolving power

To show the resolving power of the human eye draw two dots about 2 mm apart on a sheet of paper stuck to the board. View them from various distances, calculate the angle subtended by them at the eye when they appear to merge into one and also study the effect of dim light on the ability to resolve fine detail. In theory, as long as the light level is not too low, you should do better in dim light because the resolving power of a circular aperture (your pupil) depends on its diameter and in dim light the diameter of your pupil will be larger.

Theory: the smallest angle ϕ that can be resolved in light of wavelength λ by a circular aperture of diameter a is given by $\phi = 1.22\lambda/a$

Apparatus required: •Two small dots drawn close together on a piece of paper.

8 Diffraction with sound

The diffraction of sound can be shown by using the single-slit technique. The slit is an aperture about 10 cm wide between two boards. Sound of frequency about 10 kHz is emitted by a loudspeaker placed behind the gap and the diffraction can be observed by moving a microphone across in front of the gap. Feeding the output to an oscilloscope or meter can give a measure of intensity against position.

Apparatus required: •Two boards •Loudspeaker and signal generator •Oscilloscope.

9 Diffraction through tights

This is a very good example of diffraction from an irregular obstacle giving circular diffraction rings. Take a section of the tights, stretch it between your hands and then view a small torch bulb through them. Lovely coloured diffraction rings can be observed due to the diffraction of the white light from the irregular weave of the tights. Moving the tights towards and away from your eyes helps to make the coloured rings more easily visible.

Compare the effect with the regular diffraction pattern that you get from a clean finely woven handkerchief. You can produce your own obstacles for this experiment by photographing a set of irregular dots (drawing shading film is ideal) and a set of regular dots. The negatives are then used with the light. The initial size of the pattern can be altered using a computer or a photocopier to scale the image.

Apparatus required: •Small torch bulb •Tights •Handkerchief.

POLARIZATION

General theory for this section

A polarized wave is one where the vibrations are in one direction only. The human eye cannot distinguish between polarized and unpolarized light.

Malus' Law. This is an equation that gives the intensity of light I transmitted by a piece of polaroid. $I = I_{o}\cos^2\theta$ where θ is the angle between the planes of polarization of the polarizer and analyser.

Brewster's Law. $\tan p = n$ where p is the polarizing angle and n is the refractive index of the material. At this angle of incidence the reflected beam is completely plane polarized.

Polarization is a way of distinguishing between longitudinal and transverse waves: transverse waves can be polarized while longitudinal waves cannot.

1. Polaroid and photoelastic stress with the overhead projector
2. Polarization by reflection
3. Polarization of a TV signal
4. Polarization and a calculator
5. Sunset in milk in water

1 Polaroid and photoelastic stress with the overhead projector

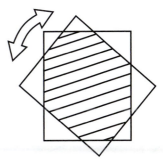

To demonstrate the effect of polarization, place two overlapping pieces of polaroid on the overhead projector (it only works using the type of projector where the light comes from underneath the transparency). Rotating one polaroid will show how the light intensity varies with angle and if this is measured with a light meter it is possible to get a verification of Malus' Law.

A beautiful extension to this is to put a piece of plastic bag or a clear plastic ruler or protractor between the two crossed polaroids to show photoelastic stress. One piece of polaroid should be placed on the glass of the projector with the plastic ruler on top of it and the second piece of polaroid put on top. The plastic rotates the plane of polarization of the light and different wavelengths are rotated different amounts by the stressed areas of the plastic giving beautiful coloured areas. A small cut in the plastic also shows localized stress patterns.

Apparatus required: •Overhead projector •Two pieces of polaroid •Plastic bag •Protractor and/or clear plastic ruler.

2 Polarization by reflection

View the glare from desks or roads through either a piece of polaroid or a pair of polaroid sunglasses. Rotation of the polaroid will cut down the glare, hence the use of polaroid glasses for driving. Reflection from a glass-fronted cupboard also shows the effect very well. The existence of a polarizing angle p, which for glass is about 57°, is easy to show. Use a TV camera if possible to show the effect to the whole class at once.

Apparatus required: •Polaroid sheets or polaroid sunglasses •Glass-fronted cupboard •TV camera if available.

3 Polarization of a TV signal

It is easy to show the plane of polarization of a TV signal by simply rotating the aerial. The outdoor type is the best with the rod and a number of half-wave bars mounted across it. Are all the TV stations in your area polarized the same way and what would happen if they weren't?

Apparatus required: •TV set •TV aerial.

4 Polarization and a calculator

The liquid crystal display on a calculator (or laptop computer) is polarized. This fact leads to yet another very impressive yet simple demonstration. Put a piece of polaroid in front of the calculator screen and just rotate it until the display disappears. Using the TV camera to show the effect to a class is especially helpful.

Apparatus required: •Calculator •Piece of polaroid •TV camera if available.

5 Sunset in milk in water

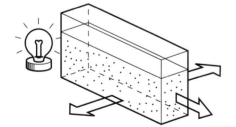

Sunsets, the scattering of light by small particles and the polarization of scattered light can all be demonstrated by the following two experiments.

(a) Fill a rectangular plastic tank about three-quarters full with water. Then add a few drops of milk. Mix the milk in and then shine a light through it (a projector is ideal). Viewed from the opposite end the lamp looks red like a sunset while when looked at from the side it looks blue, tending to green. This difference in colour is due to the tiny fat particles in the milk scattering the light, blue is scattered more than red so looking through the tank most of the blue has been scattered out leaving red light.

Testing the light scattered from the sides will show that it is also polarized! Don't add too much milk — a drop or two is enough to start with.

This explains why the sky is blue — light scattered from the particles in the atmosphere, and why the sunset is red — only the red is left after the light has passed through a large thickness of atmosphere, other colours having been scattered out. Speculate on the colour of the sky on planets with denser atmospheres than the Earth.

(b) Another version of the sunset in a tank uses slightly more specialist liquids. Make up a solution of 5 gm hypo (sodium thiosulphate) to every litre of water required. Add 15 cc (per litre) of hydrochloric acid containing 0.5 cc concentrated acid.

I gather that the colour of the iris in the eye is due to scattering. Babies are usually born with blue/grey eyes; the molecules are not linked into large chains. As they grow up the chains join and so the scattering is reduced and the iris colour turns to brown.

Theory: scattering is proportional to the fourth power of the frequency.

Apparatus required: •Rectangular plastic tank •Projector •Milk •Dropper •Polaroid sheets •Sodium thiosulphate •Hydrochloric acid.

MISCELLANEOUS WAVES, SOUND AND LIGHT

1. Tape recorders in physics
2. Sodium lamp and Young's slits
3. Grease spot photometer
4. Painting a corridor
5. Persistence of vision
6. Absorption spectrum of sodium
7. Cross and spot for the blind spot
8. Colour discs in front of a projector
9. Laser light and the 60 W bulb
10. Binocular vision
11. A shadow photometer
12. Colour subtraction
13. Singing flames and beats
14. Silhouette photographs
15. Chaotic systems
16. Phase angle
17. Ultraviolet experiments
18. P and S wave simulation
19. Sensitive flames
20. The eye
21. Coloured shadows
22. Corrugated cardboard and wave motion
23. Using a pinhole to help you read without glasses
24. Standing waves
25. Moiré fringes in a net curtain or nightie
26. Colour filters on the overhead projector
27. Mirror drawing
28. The television stroboscopic effect: slow motion waves
29. Laser model: hand

1 Tape recorders in physics

A variable speed reel-to-reel tape recorder is useful for the demonstration of the change of pitch when the tape speed is changed. Recording a note at one speed and then doubling the tape speed will increase the pitch of the note and musicians may realize that it has in fact gone up by one octave. The reverse is true of course if you slow the tape down.

Apparatus required: •Variable speed tape recorder.

2 Sodium lamp and Young's slits

The classical double slit experiment may be performed using a set of double slits and a spectrometer. Mount a pair of double slits (separation about 0.8 mm and width about 0.2 mm) in a holder on the spectrometer table. Point the collimator at a sodium discharge lamp, remove the objective lens of the telescope and adjust the eyepiece. Good interference fringes should result. Compare this result with the experiment with a laser which shows large interference fringes across the lab. The laser fringes are sharper due to the monochromatic nature of the laser light and its greater coherence.

Theory: width of the fringes $= \lambda d/D$ where d is the distance between the slits, D is the distance of the slits from the fringes and λ the wavelength of the light.

Apparatus required: •Spectrometer •Sodium lamp.

3 Grease spot photometer

It is simple to measure the brightness of a lamp, or indeed the Sun, using the old-fashioned grease spot photometer. This is simply a spot of candle wax (or oil) dropped onto a piece of paper. If the illumination of the spot is the same from both sides of the paper then the spot virtually disappears. However if you view the spot from the side where the illumination is greater then the spot will look darker than the surrounding paper. Compare a candle with a 12 V bulb, varying the brightness of the bulb, and plot a graph of power against brightness. This traditional method avoids the calibration curve of the LDR.

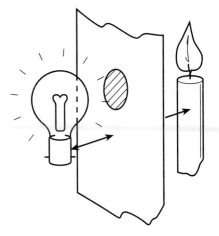

An extension of this experiment is to write a message on a large sheet of paper using melting wax. Mount the paper vertically with a light behind it. As the intensity of the light on one side in increased or decreased the writing will appear and disappear.

Apparatus required: •Prepared grease spot on a sheet of white paper in a support •Light bulb •Candle •Power supply •Voltmeter and ammeter.

4 Painting a corridor

The most reflective colour for painting a room/corridor can be investigated by using an LDR and various coloured sheets of paper. Light is simply shone on the paper and the intensity of the reflected (scattered) light measured using an LDR. For a more detailed investigation include the data sheets from RS components on the ORP12 which give the variation of resistance of the LDR with illumination in lux.

Apparatus required: •LDR •Coloured paper •Ohmmeter •Power supply •Lamp.

5 Persistence of vision

In a TV set you see 25 frames per second. The human eye is unable to detect any variation much faster than this due to persistence of vision. This can be demonstrated by using a card with a double thread fixed to each end and with a jumping horse on one side and a gate on the other. As the card is spun by tightening and slackening the threads it looks like a horse jumping a gate!

Apparatus required: •Card with horse on one side and gate on the other •Thread.

6 Absorption spectrum of sodium

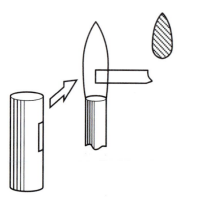

Set up a sodium discharge lamp in line with a bunsen burner and a white screen. Light the bunsen and shine light from the discharge lamp onto the bunsen flame (blue). On the screen all that is seen are some convection shadows but little else. Now hold a stick of sodium chloride in the flame.

A superb black shadow should be seen on a screen. Holding a wooden splint in the flame until it burns shows no shadow. The dark shadow produced when the sodium chloride stick is held in the flame is due to the sodium within it absorbing the photons of just the right energy from the light emitted by the sodium lamp. The bright flame is due to the emission of photons and the re-emission of the absorbed photons in all directions.

Apparatus required: •Sodium lamp •Stick of sodium chloride •Bunsen and board •Retort stand and clamp.

7 Cross and spot for the blind spot

Draw a cross and a spot on a piece of paper 12 cm apart. Hold the paper up with the cross in front of your left eye and about half a metre away. Look at the cross with the right eye while closing the left. If the paper is moved backwards and forwards a place will be found when the spot will disappear as the image of it is falling on the blind spot of the right eye.

8 Colour discs in front of a projector

Coloured objects appear different colours when viewed with light of different wavelength. This can be demonstrated by sticking three discs of coloured paper onto a board and then illuminating them with light from a projector, first using white light, and then with green, red or blue filters. This can be done either by using prepared coloured slides or by simply holding pieces of acetate colour filters in front of the projector. Relate this effect to the difficulty of buying clothes under artificial light. (if you have large pieces of film they can be mounted in card frames and placed on the overhead projector).

Apparatus required: •Projector •Sheet of paper with a disc of red, green and blue paper fixed to it •Coloured filters.

9 Laser light and the 60 W bulb

It is instructive to compare the intensity of illumination of these two lights as a source of potential danger to the eye. Consider the 60 W light bulb. Assuming that all the electrical energy is converted to light, and that it is all radiated uniformly, then at a distance of 1 m from the bulb the power density is 4.8 W m^{-2}. Compare this with the laser. Although the power of a laser of the type used in schools is only 1 mW the area of the laser beam is some 2 mm^2, remaining fairly constant with increasing distance and so giving a power density of 500 W m^{-2}, over 100 times greater.

10 Binocular vision

The slightly different appearance of a view from your two eyes that gives us the perception of depth can be shown by this simple experiment. Put your two index fingers together and look past them into the distance. The ends seem to join together to make a finger sausage!

11 A shadow photometer

This is a method of comparing the brightness of two sources of light by using the 'density' of the shadows that they give. A board marker pen is stood in front of a white screen and a lamp is set up to cast a shadow of the marker on the screen. A second lamp is placed by the first, so giving another shadow. The two lamps are equally bright when the two shadows are of equal depth and the two lamps are the same distance from the screen.

Apparatus required: •Board marker or pencil •Two lamps •Power supplies and meters.

12 Colour subtraction

Three filters, one red, one green and one blue, placed on an overhead projector and overlapped is an easy way to show the subtraction of colours.

Apparatus required: •Overhead projector •Colour filters.

13 Singing flames and beats

Draw out a 1 cm tube to give a bore with a diameter of 0.5 mm. Connect it to a gas supply, light the gas emerging from the fine end of the tube and put it in a glass tube of diameter 3.5 cm and 83 cm long. A loud noise results. Now try another with a length of tube one or two cm shorter. When both flames are lit beats are produced between the different pitched notes.

Apparatus required: •Glass tubing •Wide glass tube •Bunsen.

14 Silhouette photographs

This is an interesting introduction to photography. You will need a dark room, some photographic paper and chemicals (developer and fixer) and a series of interestingly shaped objects such as leaves, cogs, protractors and jewellery. Place the object on the paper using only the red safe light. Switch on the main light for two or three seconds. An invisible latent image will be produced on the paper, which can then be developed. All the examples suggested make excellent shadow pictures. Positives can also be produced using the technique described in geometrical optics, experiment 17.

Apparatus required: •Flat interesting shaped objects •Dark room •Photographic paper, developer and fixer.

15 Chaotic systems

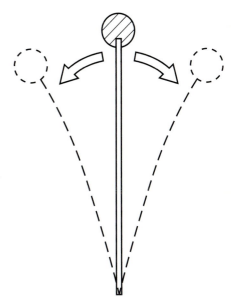

(a) Chaotic systems have a number of stable states and this can be shown by using a long flexible strip of metal rather like a giant hacksaw blade. This is mounted vertically and held by at the base by a clamp. If the strip is pulled sideways it oscillates but always comes to rest in the vertical position. If a lump of modelling clay is fixed to the top it now has three stable states — one vertical as before but there are now two more – one on either side of the vertical. You can never be sure in which of these states it will finally come to rest if disturbed from its original position (Bristol University demo).

  (b) A further demonstration of chaos and the sensitive dependence of the outcome on the initial conditions is to suspend a steel ball bearing over two magnets. When it is swung the ball bearing exhibits a chaotic path and the point where it finally comes to rest depends critically on its point of release. Doing this experiment on an overhead projector makes the motion clearly visible to a class.

Apparatus required: •Long whippy piece of steel with a holder fixed to the top •Ball bearings •Thread •Overhead projector.

16 Phase angle

Mount a ball on a rotating table and use a lamp to cast a shadow of the ball on the screen (the lamp should be a long way from the table and the table relatively close to the screen). Rotate the table and record the position of the ball and the angle through which the table has turned; the zero could be taken when the shadow of the ball is either in the centre of its traverse or at one end. Plot a graph of the displacement of the shadow against the sine (or cosine) of the angle turned by the table.

Apparatus required: •Ball •Rotating table •Motor with power supply to drive table •Projector or other suitable light source.

17 Ultraviolet experiments (safety considerations vital here)

Some very nice effects can be shown with ultraviolet light as long as the correct safety precautions are followed. Clean hair shows a greenish hue, teeth gleam (ultraviolet safety glasses are essential for this one) and fluorescein glows green. I have collected a set of other objects, mostly from gift shops: fluorescent T-shirts, silly putty (see elasticity experiment 7), rocks (the correct wavelength of ultraviolet light is needed here), a star chart, security markers, an advent calendar and so on.

18 P and S wave simulation

In an earthquake two forms of wave are transmitted through the earth – the primary or P wave, a longitudinal or push vibration, and the secondary or S wave, a transverse or shaking vibration. The differing properties of these two types of vibration can be shown by the following experiment. Use a plastic beaker with tracing paper stuck round it and with a small bulb fixed on one side. Place a beaker of water inside and observe the passage of light through the system. A bulb on one side represents the epicentre of the earthquake and the paths of light through the double beaker represent the paths of the P and S waves.

Apparatus required: •Plastic beaker •Glass beaker •Tracing paper •Small light bulb and battery.

19 Sensitive flames

The effect of sound waves on a fine flame can be shown by this experiment. Draw a glass tube out from a diameter of 1 cm to a diameter of 1 mm. Connect it to the gas supply and light the end of the tube — a long fine flame should be produced. This can now be used to investigate sound levels such as nodes and antinodes in standing waves. Modern versions use a sound level meter but are perhaps not so visually impressive.

Apparatus required: •Glass tube •Gas supply.

20 The eye

The following series of simple experiments can be performed to study the physics of the eye.
   (a) The ability to resolve fine detail: draw two dots about a millimetre apart on a piece of paper, fix this to a wall and see how far away students have to stand until they can only see it as one dot.
   (b) Look at newspaper pictures and actual photographs using a lens and compare the two.
   (c) Help with a lens: see how the use of a lens enables you to read print very close up.
   (d) Optical illusions: just for interest.

21 Coloured shadows

The addition of colours can be seen clearly if light from three ray boxes each containing a different coloured filter (red, blue or green) is shone onto a white sheet of paper. Varying the intensity of each will give any colour of the spectrum. Standing three board markers in the way gives three superb coloured shadows, each being the sum of the colours from only two of the ray boxes.

Apparatus required: •Three board writers •Three ray boxes and power supplies •Three coloured filters.

22 Corrugated cardboard and wave motion

Using a square of corrugated cardboard or plastic which has been cut at an angle to the corrugations gives two axes of sine waves. This can be used as an aid when teaching the wave equation $y = A\sin2\pi(ft - x/\lambda)$.

The wavelike corrugated edge on one side of the square gives the x variation and the other at right angles to it the t variation while the height at any point represents the y value or displacement of the wave.

Theory: $y = A\sin2\pi(ft - x/\lambda)$; one side gives varying t for fixed x (say $x = 0$), $y = A\sin2\pi ft$, and the other varying x for fixed t (say $t = 0$), $y = A\sin2\pi x/\lambda$.

Apparatus required: •Sheet of corrugated plastic or cardboard (plastic is better — the corrugations can be bigger).

23 Using a pinhole to help you read without glasses

This is an amazingly simple experiment but of great help to anyone who forgets their glasses! Make a pinhole and look through it at print — you can read it without glasses! This works for both long- and shortsighted-people. An alternative is to look through the small hole made between the tips of your first and second finger and the side of your thumb. The advantage of this method is that you always carry your fingers with you and that the size of the hole is variable.

The phenomenon is explained by understanding that as the aperture is reduced the depth of focus is increased.

Apparatus required: •Pinhole in a piece of card.

24 Standing waves

The effect of the length of a cord rather than its tension on the standing waves that can be produced on it can be shown very simply by this experiment. Fix the end of a string to an off-centre hole in a wheel on an electric motor (or the top of a vibration generator). Hold the other end but have it passing through a short length of plastic tube or bung to allow it to rotate freely. Switch on and move your hand towards and away from the motor. Large standing waves result.

Theory: frequency of the fundamental standing wave on a cord of length L, tension T and mass per unit length m is given by the equation $f = \frac{1}{2}L(T/m)^{1/2}$ so the frequency is inversely proportional to the length of the cord.

Apparatus required: •Cord •Motor and power supply or vibration generator and signal generator.

25 Moiré fringes in a net curtain or nightie

The overlapping of the fine weave of two net curtains or two layers of a fine nightie show good moiré fringes. Interference can also be simulated by two plastic discs with circles drawn on them placed on the overhead projector.

Apparatus required: •Net curtain or nightie •Overhead projector (optional).

26 Colour filters on the overhead projector

Show disappearing writing by shining light through a filter on the overhead projector overlaying a transparency. For example if a red filter is used beneath some red writing the writing should disappear if a suitable shade of red is used.

27 Mirror drawing

Eye–hand co-ordination can be demonstrated using the following irritating experiment. Draw two five pointed stars on a piece of paper, one inside the other, and then try to draw between the outlines when you view them using a plane mirror.

Apparatus required: •Plane mirror •Paper •Pencil •Sheet of card to obscure direct view.

28 Stroboscopic effect with a television

A wonderful example of both the stroboscopic effect and the motion of
waves on a stretched string can be obtained using a television and a
rubber band. Switch on the television and if possible tune to a static clear
screen or else connect a TV camera to the television and point the camera
to a blank board. Now stretch the rubber band in front of the screen and
pluck it. You will see waves travelling slowly along the band — the
scanning property of the television acts like a stroboscope. By suitable
experimenting with the length and tension of the band the waves can be
made almost stationary. I have tried both suspending a weight from the
band and also using a screw thread to vary the tension.

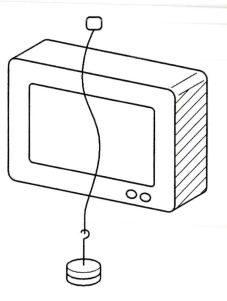

Apparatus required: •Television •Rubber band •Set of slotted masses
•Retort stand •TV camera if possible.

29 Laser model: hand

The stimulated emission from a laser can be simulated as follows.
Suspend a piece of card from a rod by two threads with a pendulum bob
on the end. Get an electric fan and blow a jet of air at the card. Now move
your other hand up and down in the air stream. If you get the frequency
right the oscillations of the card will build up in just the same way as a
beam of light along the axis of a laser increases in intensity as it reflects
backwards and forwards between the ends of the tube.

Apparatus required: •Card •Hair dryer or fan •Thread •Rod •Pendulum
bob.

EXPANSION OF SOLIDS AND LIQUIDS

1. Expansion of metal

2. Bi-metalllic strip: heating and cooling

3. Iron rod in projector: expansion of metals

4. Expansion of a liquid

5. Jumping metal discs

6. The expansion and contraction of glass

1 Expansion of metal

Hold a long strip of aluminium cooking foil horizontally and tightly between two clamps. Heat it from beneath with some candles; significant sag can be produced. (Compare this with the hot wire ammeter; see Current Electricity, experiment 4.)

A thought-provoking question is the washer problem. If you take a metal washer and heat it does the hole in the centre get bigger or smaller? It gets larger since all parts of the metal expand but those parts nearest the centre expand least.

Uses and effects of the expansion of solids include: rivets, telephone wires, railway lines, buildings, bridges, clock pendulums, tarmac-filled gaps in concrete motorways.

Apparatus required: •Two retort stands and clamps •Candles or bunsens •Strips of aluminium foil.

2 Bi-metallic strip: heating and cooling

A bi-metallic strip made of two metals welded together is an excellent way of demonstrating the different coefficients of thermal expansion of different metals. It will bend when heated with the metal that expands the most (brass in the case of a brass–iron strip) on the outside of the curve. An alternative demonstration with the bi-metallic strip is to put it in a fridge (or in a cooling mixture of ice and water). The strip will bend with the brass layer on the inside, showing that brass also contracts more than iron. Point out that it is the *change* of temperature that matters.

Apparatus required: •Bimetallic strip in holder or a pair of tongs •Bunsen.

3 Iron rod in projector: expansion of metals

A very simple way to demonstrate the expansion of metals is to put a steel rod in a projector beam (brass or copper will do as well of course). You will need the old type of projector and the end of the rod should be placed where the slide carrier was. The projector is switched on so that it projects a magnified shadow of the rod onto a screen a few metres away. Make a mark on the screen at the end of the shadow. Heat the rod and actually watch it expand by observing the movement of the end of the shadow.

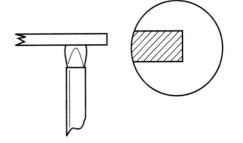

This can be used as a simple demonstration, or, if more detail is needed, measurements can be made of the magnification of the projector (width of shadow/width of rod) and a guess made at the average temperature rise of the rod to give the coefficient of linear expansion of the metal. Also show that when the rod cools down it will return to its original length (make sure nobody knocks into the bench!)

Theory: expansion of a metal bar of length L heated by θ °C $= L\alpha\theta$ where α is the coefficient of linear expansion of the metal (about 10^{-5} for most metals).

Apparatus required: •Projector •Metal rod •Retort stand and clamp •Ruler •Bunsen.

4 Expansion of a liquid

Take a round-bottomed flask and fill it to the brim with coloured water. Take a 1–2 m long piece of capillary tubing, push one end into a rubber bung and push the bung into the top of the flask. Observe the expansion of the liquid when the flask is heated. If you watch carefully it is interesting to see that initially the level of the liquid falls. The glass expands first but since it is a bad conductor of heat it takes a while for the heat to pass through to the liquid, raise its temperature and so cause it to expand.

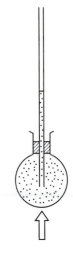

Apparatus required: •Round-bottomed flask •Capillary tubing •Bung •Bunsen •Tripod •Retort stand and clamp •Heat-resistant mat.

5 Jumping metal discs

These bimetallic discs were described in *Physics Education* (volume 22, no. 3, 1987). If you press the silvery back convex face it clicks so that this face becomes the inner or concave face. This configuration is unstable and returns to the original shape. However if the disc is warmed by rubbing it between your fingers or putting it in hot water it will remain stable in the new shape for a while. If it is now left on a flat surface it will flex and jump in the air as it cools down.

(The discs were bimetallic in structure – one face being nickel and the other stainless steel.)

---

6   The expansion and contraction of glass

In a desert the temperature at night falls rapidly as the sun goes down; the outside of the rocks cool and since the inside is still warm (and large) the outer layers attempt to contract but flake off. This can be shown by experiments with hot glass and cold water.

(a) Heat a glass rod in a bunsen flame and then plunge it into a beaker of cold water — the glass shatters.

(b) Heat a glass marble in a bunsen flame using a pair of tongs and then drop it into a beaker of cold water — the marble shatters. It falls below the water surface before shattering!

 (Safety — use a safety screen and wear safety goggles when carrying out these two experiments

Apparatus required: •Glass rod •Beaker of cold water •Safety screen and goggles •Marbles •Tongs •Bunsen.

EXPANSION OF GASES

General theory for this section

When a gas is compressed or expanded without its temperature changing (isothermal) the product of its pressure P and volume V is constant, i.e. $P_1V_1 = P_2V_2$. This is Boyle's Law.

   To keep the temperature constant heat energy must be added during expansion and removed during compression.

   The equation that describes the behaviour of a gas under any change is the ideal gas equation: $PV = nRT$ where n is the number of moles of the gas, R the gas constant and T the absolute temperature ($T = 273 +$ temperature in centigrade).

If the volume is kept constant the pressure of the gas is directly proportional to the absolute temperature and if the pressure is kept constant the volume is directly proportional to the absolute temperature.

1. The expansion of gases: air
2. Exploding balloon: heat
3. Soap film: gas expansion
4. Popcorn and expansion of a gas
5. Expansion cooling
6. Tin can above a bunsen
7. The bicycle pump: adiabatic and isothermal changes

---

1 Expansion of gases: air

The expansion of air can be used to make a simple yet accurate air thermometer. Take a large round-bottomed flask and fill it about one third full with coloured water. Next fit a long (at least 1.5 m) glass tube into a rubber bung, so that the bung is about ten centimetres from one end. Push the bung in firmly, making sure that the end dips into the water in the flask and so about two thirds of the flask is trapped air. Heat the flask either using your hands or very gently with a bunsen. The expansion of the air in the flask forces the coloured water up the tube.

   The heat of your hands is usually sufficient to show a large expansion, and it makes an interesting and sensitive comparison between the temperature of various students' hands. Calibrating the apparatus enables it to be used as an actual thermometer.

Apparatus required: •Large round-bottomed flask •Bunsen burner •Large bore capillary tube (to make the expansion easily visible).

2   Exploding balloon: heat

Blow up a balloon, close the open end by tying it up and then heat it gently above a bunsen flame. Holding it well above the flame will prevent the rubber simply melting. The expanding air in the balloon will demonstrate the expansion of gases, bursting the balloon.

Apparatus required: •Balloon •Bunsen •Safety screen.

3   Soap film: gas expansion

An alternative method for demonstrating the expansion of air uses some delightfully simple apparatus. Use a large round-bottomed flask and make a soap film over the neck by dipping the neck into some strong soap solution. Warm the flask with your hands; the expansion of the air in the flask will give a good bubble on the end. Warming it over a bunsen flame will make it happen quicker but it is then quite easy to pop the bubble. As the air cools the bubble collapses back to a flat film.

Apparatus required: •Large round bottomed flask •Soap solution.

4   Popcorn and expansion of a gas

Use the expansion of popcorn to demonstrate the increase in the volume of air when it is heated. I usually make it in a tall 1 litre beaker covered by a saucer so that the students can see what is going on. Put enough oil in to just cover the base of the beaker, sprinkle in a layer of popcorn and then heat it gently over a bunsen. The popcorn expands in a few minutes after reaching a sufficiently high temperature.

Cleaning the beaker thoroughly beforehand enables them to take the resulting popcorn away to eat later — a popular way to end the lesson!

Apparatus required: •Popcorn (uncooked) •Tall 1 litre beaker •Saucer to cover the beaker •Cooking oil •Bunsen •Tripod •Gauze •Heat-resistant mat •Sugar (to taste!)

5   Expansion cooling

The formation of ice round a gas cylinder when the gas escapes is a very good demonstration of cooling on expansion. A very impressive demonstration of this effect is to show the formation of dry ice as gas expands from a high-pressure carbon dioxide cylinder. Even the small carbon dioxide filled bulbs used in a soda siphon will show this. They can be punctured with care using a compass point and show a marked cooling as the gas emerges. Some idea of the energy transfer on expansion can be gained by holding the bulb in a beaker of water using a pair of wooden test tube tongs.

Safety goggles must be worn when carrying out this experiment.

Apparatus required: •Carbon dioxide cylinder and cloth •Soda siphon bulb •Sharp point to puncture the bulb (a pair of compasses or dividers is ideal).

6   Tin can above a bunsen.

The expansion of a gas as its temperature rises can be shown quite impressively by the following experiment. Get a small metal tin with a tightly fitting metal lid. Put it on a tripod behind a safety screen and heat the tin. After a few moments the expansion of the air in the tin should blow the lid off. (Don't put the lid on too tightly and don't go to check if it hasn't worked — turn off the gas and wait for it to cool down).

Tip: a little water in the tin will help things along and ensure a good explosion although some students may notice the steam. Use a safety screen.

Apparatus required: •Metal tin with a tight fitting lid •Bunsen •Safety screen •Tripod •Heat-resistant mat.

7   Bicycle pump: adiabatic and isothermal changes

The rapid compression (or expansion with the valve reversed) of the air in a bicycle pump shows adiabatic changes. Normal pumping heats up the air in the pump while a rapid expansion shows cooling. A slow change allows heat transfer and so there is no change of temperature — an isothermal change.

Apparatus required: •Bicycle pump •Thermometer or thermistor.

CONDUCTION

General theory for this section

Conduction occurs by the transfer of energy through a material by the collisions of molecules. In metals conduction is further increased by the free electrons carrying energy from places of high temperature to those of lower temperature.

1. Espresso coffee/or beer: conduction in foam
2. Baked Alaska
3. Sweaters and anoraks: the thermal conductivity of air
4. Thermal conductivity of water
5. Thermochromic paint for conductivity
6. Line of students: conduction
7. Lifting up a flame and copper gauze on chip pan: conduction in gauze
8. Conductivity of air
9. Wooden rod and copper tube

1   Espresso coffee/or beer: conduction in foam

My wife and I were sitting on Waterloo station, about to drink a cup of really frothy espresso coffee when she looked at the foam on the top and said 'I suppose that helps to keep it hot!'

Indeed, the thick layer of foam on a cup of espresso coffee helps to insulate the liquid below it. There is a considerable amount of air in the foam, which makes an insulating blanket.

An extension of this would be to investigate the cooling of the same mass of frothy coffee and non-frothy coffee – both starting from the same initial temperature in the same sized beakers.

Apparatus required: •Frothy coffee •Non-frothy coffee •Beakers •Thermometers.

2   Baked Alaska

This tasty dessert shows the insulation effects of the air within the meringue. As the meringue cooks the air pockets within it form an insulating blanket and the ice cream within it does not get hot enough to melt as long as the cooking is fairly rapid.

3 Sweaters and anoraks: the thermal conductivity of air

As a demonstration of the poor thermal conductivity of air get a student to put on as many layers of clothing as they can. (Small students are better — they can wear more layers of adult clothes!) Use sweaters, anoraks, pullovers, jackets and a lab coat. Don't let them wear them too long — it gets really hot inside! You could use a thermometer probe to investigate the temperature beneath the layers of clothing.

Apparatus required: •Lots of large thick clothing.

4 Thermal conductivity of water

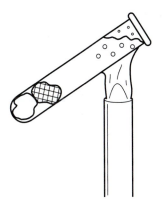

The relatively poor thermal conductivity of water can be shown by putting a piece of ice in the bottom of a test tube and holding it in place with a piece of copper gauze wedged above it in the tube. Fill the rest of the test tube with water and heat it at the top until it boils. Due to the poor conductivity of water the ice at the bottom remains solid even though the water is boiling at the top.

Apparatus required: •Piece of ice •Test tube •Water •Bunsen •Heat-resistant mat •Tongs •Metal gauze.

5 Thermochromic paint for conductivity

This specialist paint can be used to paint different metal rods in a variation of the simple conductivity experiment. In the simple experiment two rods of different metals are held in a bunsen flame and their conductivity is compared by the students seeing which one they have to put down first before it gets too hot. This one has the highest thermal conductivity. To avoid burns simply paint the rods with the thermochromic paint, hold them in a non-conducting (wooden) clamp and watch the change in colour of the paint.

   Another version is to soak some filter paper in a solution of cobalt chloride and allow it to dry. The paper is now heat sensitive and will turn blue as it gets hot. Place a piece of the treated paper on a tripod; the rods from the earlier experiment can then simply be rested on it while their other ends are heated with a roaring bunsen flame. The rate at which the paper in contact with the rods turns blue gives a good comparison of their thermal conductivity.

Apparatus required: •Conductivity rods kit (rods of brass, aluminium, copper, iron, zinc and glass) •Bunsen •Tripod •Heat-resistant mats (2) •Thermochromic paint •Filter paper •Cobalt chloride solution.

6   Line of students: conduction

This is a very simple old idea to show conduction along a solid rod; transfer of energy occurs by the vibration of molecules. Each molecule stays where it is and simply vibrates — it is the energy that is transferred down the rod. The students link arms and you shake the line at one end. The vibration of the pupil at the end is transmitted along the line until the one at the other end is moving. Don't shake them so hard that they fall over! This simple demonstration refers to a non-metal solid. The ingenious teacher might be able to devise a way of showing conduction in a metal where much of the energy transfer is due to free electrons. If you are really brave take the group of pupils to the gym or sports hall and get them to kick a large number of footballs about — not recommended in the physics lab!

7   Lifting up a flame and copper gauze on chip pan: conduction in gauze

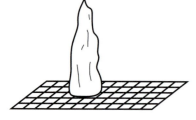

This experiment clearly shows the high thermal conductivity of copper and also the need for gas to be at a sufficiently high temperature before it will burn. Light a bunsen and hold a piece of copper gauze over the outlet, resting on top of the bunsen tube — the flame burns through it. Then gently lift the gauze — the flame comes up with it, burning above the gauze and leaving an area of unburnt cooler gases below it. The heat is conducted away from the gauze and into your hands, leaving the gases below the gauze too cold to catch fire. It is a very good demonstration of how the traditional Davy safety lamp works. It is possible to lift the flame right off, thus putting out the gas. Don't forget to turn it off afterwards!

As a practical application of this a sheet of copper gauze can be placed over a chip pan to prevent the fat catching fire. This is another example of the conduction in metals.

Apparatus required: •Bunsen or candle •Piece of copper gauze •Heat-resistant mat.

8   Conductivity of air

Hold your hands carefully either side of a roaring bunsen flame – you feel virtually no heat since the transfer of heat from the flame to your hands in this position is only due to the conduction of heat through the air. This demonstrates very simply that air is a very poor conductor of heat. It feels much hotter if they are held above the flame showing the principles of convection: air expands when it is heated, its density goes down and so the hot air rises!

Apparatus required: •Bunsen or candle •Heat-resistant mat.

9 Wooden rod and copper tube

This is a very good demonstration of differing thermal conductivities. The apparatus consists of a copper tube, one end of which fits tightly round a wooden rod. A piece of paper is wrapped tightly around the join of the wood and copper and the paper over the join is heated gently in a bunsen using a clear blue flame. The rod should be rotated during heating to stop the paper catching fire. The paper will blacken over the poorly conducting wood while staying undamaged over the better conducting copper. A very good comparison between the conductivities of two materials can be made by walking in bare feet on a carpet and then on a tiled floor. Although both are at the same temperature it feels colder on the tiled floor since the tiles are a better conductor than the carpet and so heat energy is transferred from your feet to the floor.

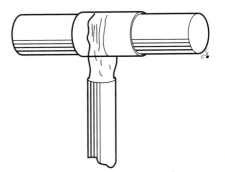

Apparatus required: •Bunsen •Heat-resistant mat •Composite wood/copper rod.

CONVECTION

General theory for this section

This section deals with convection currents in fluids – a fluid can be either a liquid or a gas.

Most people remember the phrase 'Hot air rises' but why does it?

Convection currents occur because a fluid expands when it is heated. This expansion reduces its density so the low density fluid will rise through that of higher density. Convection is therefore the transfer of heat energy by the movement of the more energetic molecules (the 'hotter' ones) from one place to another.

1. Christmas table decoration: convection currents
2. Convection: a cardboard serpent
3. Convection in a flame
4. Heat loss from young animals: the effect of surface area
5. Convection currents
6. Falling candle
7. Lava lamp
8. Convection in the high chimney and the double chimney

1   Christmas table decoration: convection currents

This uses a lovely wooden table decoration that can be bought widely. Around the base are a set of candles and at the top is a wooden rotor with a series of large angled blades. The rotor is fixed to part of the decoration including figures and animals. Hot air rising from the set of candles turns the rotor which rotates the whole display.

Apparatus required: •Table decoration and candles •Matches.

2   Convection: a cardboard serpent

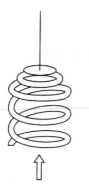

Cut a paper, card or foil snake-like coil and suspend it by a thread above a light bulb. The rising hot air will rotate the coil giving a simple but effective demonstration of convection in a gas. The sequence is: heat energy expands the air, thus lowering its density; this lower density air then rises.

Apparatus required: •Paper serpent •Thread •Stand •Light bulb in holder •Card. (I have found that a circle of card about 15 cm across cut into a spiral works well if you cut it into a spiral with arms about 1.5 cm across leaving a disc in the middle to attach the thread.)

3 Convection in a flame

The convection currents in a flame can be seen by casting a shadow of it onto a screen. Candles and bunsens with a blue non-roaring flame work well using a projector to show the movement of the low density air above the flame.

Apparatus required: •Bunsen •Heat-resistant mat •Projector.

4 Heat loss from young animals: the effect of surface area

We used to estimate the surface area of a small mammal by wrapping it in a paper tube — maybe not suitable these days but we can still compare the surface area to mass ratio for young and old animals. You can extend this to the calculation of surface area to mass ratio. For example, a sphere has the smallest surface area to mass ratio for a given material. A set of wooden blocks can be made into a tall thin person or a short fat one – the tall thin one has the greatest surface area and would therefore lose heat most rapidly from the surface. The surface area of a pupil can be found approximately by wrapping them in sheets of newspaper.

Try boiling some potatoes and measuring how rapidly they cool, some being cut into small pieces. The smaller ones have the greatest surface area to mass ratio and so cool the quickest.

Apparatus required: •Wooden blocks •Saucepan •Potatoes •Thermometers •Newspapers.

5 Convection currents

Convection currents in liquids can be easily shown with potassium permanganate crystals in water. You can do this in a beaker of water or use the special rectangular test tube available from the manufacturers. With the beaker simply drop some crystals into the water near to one side of the beaker. They will fall to the bottom, but if the beaker is now heated at the base on the opposite side the colour will be drawn across and rise up that side.

With the rectangular tube drop a few crystals in the top and then heat one bottom corner. The crystals dissolve and the colour moves round the tube showing the convection currents in the water. Catching the moving colour by moving the bunsen enables you to change its direction.

A suspension of aluminium paint in water is an alternative technique

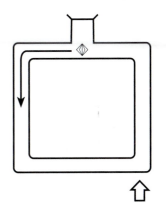

and more long lasting as with the potassium permanganate the water simply turns pink after a while and no convection can be observed.

Apparatus required: •Beaker or rectangular test tube •Bunsen •Potassium permanganate crystals •Heat-resistant mat.

---

6 Falling candle

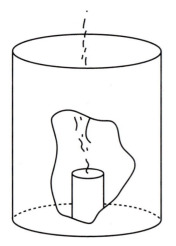

This intriguing demonstration shows what happens when the effects of convection are 'cancelled out'. A candle is fitted to the inside base of a tin; the candle is now lit and the tin is dropped. The tin should be big enough to provide enough oxygen, but the flame still goes out.

Theory: Since there is effectively zero g in the can one of the conditions for convection does not apply — that is, although the hot waste gases in the candle flame have expanded they do not rise away from the flame since they and the flame are both falling. This means that there is no convection in the can. The candle therefore tries to burn in its own waste products and fails. Helen Sharman tells of how she had to sleep in a draught from a fan in the Russian space craft so that she didn't suffocate in her own exhaled breath. Michael Foale had a similar experience in Mir.

Apparatus required: •Tin can •Candle •Matches •Protection for the floor (a piece of old carpet) •TV camera if available.

---

7 Lava lamp

This is a commercial lamp. As it warms up the material in the liquid rises and falls showing beautiful convection currents, rather like Galileo's thermometer in some ways. It would be very nice if the experiment with aniline in salty water could still be done but I understand that aniline has possible carcinogenic effects and so should be avoided. (See Density, upthrust and Archimedes, experiment 9.)

Apparatus required: •Lava lamp

---

8 Convection in the high chimney and the double chimney

(a) This first experiment shows the effect of convection in a high chimney. Use a tall metal tube to show the convection currents in a chimney. (I use a piece of metal pipe about a metre and a half long and some 10 cm in diameter). A bunsen should be placed at the bottom of the chimney and if small pieces of paper are introduced into the flame they will be shot out of the top of the chimney by the rising air currents. It always reminds me of when I put our first Christmas tree onto the fire in our flat at the bottom of a four storey block! The convection currents were enormous and it 'went up like a torch' – not to be recommended!

(b) Another nice demonstration uses the commercially available double chimney — simply a glass fronted box with two glass chimneys in the top with a candle below one of them. If the candle is lit convection is created; hot air rises out of one side and draws cold air in at the other. A smouldering taper held over the chimney without the candle will show the smoke being drawn down into that side and ejected at the other by the hot air currents.

A further variation of this is to put a candle on the bench, light it and then lower a large glass tube (half a metre long and at least 6 cm in diameter) down over it so that the base of the tube rests on the bench. The candle soon goes out due to lack of oxygen. Repeat the experiment but this time lower a metal plate down the centre of the tube so that it hangs just above the candle flame. The flame stays alight; hot gases escape up one side of the plate and 'fresh air' is drawn down the other.

Apparatus required: •Metal chimney •Retort stand and clamp •Double chimney apparatus •Candles •Bunsen •Matches.

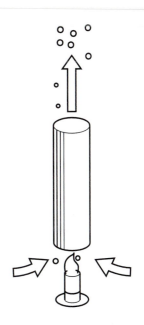

RADIATION

General theory for this section

The range of infrared radiation is from roughly 750 nm to 10 000 nm (7.5×10^{-7} m to 10^{-5} m or one hundredth of a millimetre). Long-wave heat radiation will not pass through glass. Hot metal objects will start to glow when they reach a temperature of around 500°C.

Black surfaces emit and absorb heat radiation better than shiny white ones. Shiny surfaces are the best reflectors.

1. Greenhouse effect
2. Crookes' radiometer
3. Infrared radiation: the TV/stereo remote controller
4. Radiation from thermometers
5. Thermoscope: radiation detector

1   Greenhouse effect

For this simulation you need a clear plastic sandwich box, two thermometers and a sunny day. Place one thermometer in the box and the other in the sun on the ground outside it, preferably sheltered from draughts. Leave them for an hour and observe the effects on the thermometers.

Theory: the greenhouse effect occurs because the radiation from the Sun (surface temperature 6000°C) has a spread of wavelengths which peak in intensity at around 500 nm and this radiation can penetrate clouds of methane, water vapour, carbon dioxide, nitrous oxide and CFCs in the atmosphere. However, when the solar radiation falls on the ground it only raises its temperature to around 20°C. The radiation emitted from an object at this temperature has a peak of around 12 mm (12 000 nm) and this longer wavelength cannot penetrate the gaseous clouds, or in this case the plastic. The air below therefore heats up!

Apparatus required: •Sandwich box and two thermometers •A radiant heater if there is no direct sunlight.

2   Crookes' radiometer

This is an excellent piece of apparatus for demonstrating that dull black surfaces absorb heat radiation better than shiny ones. The glass bulb is filled with low pressure air and as radiation falls on the vanes the black surfaces absorb more radiation than the silvered ones, the air near them heats up, expands and so pushes the vanes round.

Try placing a sheet of glass or a beaker of water between the heat source (a small electric fire, candle, light bulb or bunsen burner) and the radiometer and investigate the effect of these materials on the transmitted infrared radiation.

Apparatus required: •Crookes' radiometer •Heat source •Glass •Beaker of water.

---

3 Infrared radiation: the TV/stereo remote controller

In the TV remote control device we have an ideal focused beam of infrared radiation. This can be used to study:

(a) the absorption of infrared by glass and plastic

(b) the reflection of infrared from flat and rough surfaces

(c) the diffraction of infrared through a narrow slit — it is left for the reader to decide what would be a 'narrow slit' for infrared radiation!

The best detector that I know for infrared is a small TV camera. Connect the camera to a colour TV receiver and point the TV remote control towards it, focusing the camera to give a clear image on the screen. Now press any of the buttons on the remote controller. Although you cannot see any radiation emitted by the controller the camera will pick up the infrared and a bright flash will appear on the TV screen! My 18-year-old students have queried the colour of this light!

Apparatus required: •TV remote controller •Glass •Plastic •Metal sheet •Narrow slit •TV camera and television.

---

4 Radiation from thermometers

The different amount of absorption of heat radiation by different surfaces can be studied using the following simple experiment. Take two thermometers, blacken the bulb of one of them with soot or black paint and then place them both in boiling water. When the temperatures of both have reached 100°C (or as near as possible, depending on the boiling point of water that day) take them out quickly and fix them in clamps in the air. Observe the rate of cooling. The blackened one will cool much more rapidly showing that black surfaces emit heat better than shiny ones.

The absorption of heat by black surfaces is very obvious when walking round a swimming pool on tiles. The darker coloured ones absorb more heat energy from the Sun and so get hotter — they can be too hot to step on in bare feet!

Apparatus required: •Thermometers, one blacked with soot or black paint •Boiling water •Clamp stands.

5   Thermoscope – radiation detector

This is made from two glass bulbs (round flasks will do) joined by glass and rubber tubing and partly filled with water. One flask is blackened, the other is silvered with aluminium paint and a lamp is placed between them. Due to the different amounts of absorption of radiation of the two flasks one gains more heat energy than the other, the vapour pressure inside the black one increases more rapidly and water moves along the glass tube from the black to the shiny bulb. It used to be possible to buy them with ether inside and this works much better because of the volatile nature of ether, but it is probably not very safe to make one up in the lab.

SPECIFIC/LATENT HEAT

General theory for this section

When heat energy is applied to an object its temperature rises.

Heat energy = mass m × specific heat capacity c × change in temperature θ (the energy is in joules if the mass is in kg, the temperature in °C and the specific heat capacity in J kg^{-1} °C^{-1}).

When a substance changes its state, energy is needed to do this.

Specific latent heat of vaporization (heat needed to change 1 kg of the liquid into vapour). Specific latent heat of fusion (heat energy needed to change the state of 1 kg of solid into a liquid). Using an electrical heater:

Energy input = power × time = $V I t = (V^2/R)t$.

Specific heat capacity

1. Energy in a candle
2. When do you add the milk?
3. Heat energy in a flame
4. The most effective immersion heater
5. Power of a bunsen burner
6. Steamed pudding and specific heat capacity

Latent heat, melting, boiling and evaporating

7. Evaporation
8. Milk bottle top in cold weather
9. Latent heat: kettle
10. The freezing mixture
11. Regelation and the ice block
12. Boiling water under reduced pressure
13. Change of volume of melting ice: cast iron flask
14. Change of volume of melting ice: burette
15. Boiling under reduced pressure: an alternative method
16. Fish in a pond
17. Floating ice cube

1   Energy in a candle

An estimate of the heat energy given off by a candle can be gained by the following experiment. Use the candle to heat about 250 g of water in an aluminium calorimeter and measure the rise in temperature. Hence work out the energy given off by the length of candle burned. Then by measuring the length of a whole candle work out the energy produced by a whole candle and finally the comparative cost of a kW h of energy produced by using either the mains electricity to drive a heater or the candle. The mains is much cheaper by a factor of up to ten!

(Specific heat of water: 4200 J kg^{-1} °C^{-1}; aluminium: 1000 J kg^{-1} °C^{-1}.)

Apparatus required: •Candle •Calorimeter •Aluminium foil •Balance.

2   When do you add the milk?

This is the classic cooling experiment. You have a hot cup of coffee and a given amount of cool milk. The problem is to decide when to add the milk so that the resulting mixture will reach a given lower temperature the quickest.

Theory: according to Newton's law of cooling, the rate of loss of heat from an object is directly proportional to the temperature difference between it and its surroundings, so the higher the temperature the faster will the coffee cool. Adding milk will immediately reduce the temperature so it is better to allow the coffee to stand for a little while before adding the milk, thus utilizing this rapid rate of cooling when the temperature difference is high.

Apparatus required: •Beaker of hot water •Beaker of cold water •Thermometer •Stop clock.

3   Heat energy in a flame

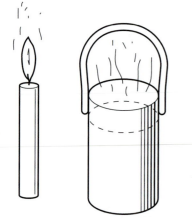

Students often have difficulty separating the ideas of heat energy and temperature and this experiment is designed to help! Compare the energy in a candle flame with the energy in a hot bath. Put out a match flame with wet fingers and then try putting your hand into a bucket of hot water. Although the flame is much hotter than the bucket of water (its temperature is about 800°C compared with the 60°C of the water in the bucket) the energy content is much less. Your hand warms up a little and the flame cools and since your hand is so much more massive than the flame there is little temperature rise.

(Safety: it needs to be done quickly otherwise a small area of your hand does heat up considerably!)

Theory: if we assume that the flame has a mass of 1 mg, a specific heat capacity of 2000 J kg^{-1} K^{-1} and a temperature of 820°C then the heat energy emitted when it cools to room temperature is just under 2 J. However, a bucket of 4 kg of hot water initially at 60°C will release 672 kJ when it cools to room temperature!

Apparatus required: •Candle •Bucket of hot water (60°C).

4   The most effective immersion heater

Use different types and dimensions of wire to heat up about 75 g of water in a 100 ml beaker of water. Using a constant voltage you can show that the low resistance wires, i.e. ones that are made of wires that are short and thick, are the best heaters. This makes a very good piece of investigative practical work.

Theory: power of heater $= VI = V^2/R$ so for a given voltage the output power is directly proportional to the current and so inversely proportional to the resistance, wires with low resistance giving more power than those of high resistance. Resistivity $\rho = RA/L$ and so power $= V^2 A/L\rho$. So the energy output in a time t is $V^2 At/L\rho$, but this is equal to $mc\theta$ and so the rise in temperature (θ) is $V^2 At/L\rho mc$.

Apparatus required: •Ammeter •Voltmeter •Wires of different types and diameters •Beaker •Power supply •Stop clock.

5   Power of a bunsen burner

This experiment acts as a useful introduction to work on specific heat capacity. Heat a known amount of water (say about 1.5 kg) in an aluminium saucepan for a known time (three to five minutes) and measure the temperature rise. Knowing the amount of energy delivered in a certain time you can then work out the power of the bunsen burner. Assume that the specific heat capacities of the water and material of the container are known.
    The powers that my pupils obtained ranged from 0.3 to 0.5 kW depending on the type of bunsen and the time of heating since they did not allow for heat loss.

Theory: energy supplied by bunsen = power × time = [(mass of water × specific heat capacity of water) + (mass of container × specific heat capacity of container)] × temperature rise.

Apparatus required: •Bunsen •Aluminium saucepan •Stop clock •Balance

6 Steamed pudding and specific heat capacity

This experiment is designed to test the statement that the jam in a steamed pudding is always hotter than the pudding. The way to do this is to actually cook a steamed pudding in its tin by heating it in a saucepan of boiling water in the lab. Use thermometers to measure the temperature of the pudding and jam. They both start off at the same temperature (roughly 100°C) when heating ceases but after a minute there could be as much as a 10°C difference between their temperatures. In fact in one experiment after getting the pudding out onto a plate and putting the thermometers in place the jam registered 70°C and the centre of the pudding 55°C. The difference in temperature is due to the much higher specific heat capacity of jam which therefore cools much more slowly. Opening the can at the wrong end can be disastrous and demonstrates the pressure effect of gases — blowing out scalding jam across the lab! (I know — I have done it!) An extension of this experiment could be to measure the specific heat of jam by heating a known mass of jam with an immersion heater. Note: jam puddings are better than syrup ones for this experiment.

Apparatus required: •Tinned steamed pudding •Two thermometers •Saucepan •Tripod •Bunsen •Heat-resistant mat.

7 Evaporation

This experiment is a simple demonstration of cooling by evaporation. Put a small drop of perfume or methylated spirits on the skin (check for allergic reactions first!). As the liquid evaporates it takes heat energy from the hand and so you feel cold. This would also work by putting the spirit on a thermometer or a thermometer probe, thus avoiding the safety angle but perhaps not so memorable.

Apparatus required: •Perfume •Methylated spirits •Thermometer (optional).

8 Milk bottle top in cold weather

A bottle of milk left out in the doorstep on a freezing cold morning soon freezes, the frozen milk pushing up the cap. You can simulate this by fitting a piece of aluminium foil over a milk bottle tightly full of water in the freezer (a rubber band helps to keep it in place). The piece of foil will bulge upwards on freezing.

Apparatus required: •Milk bottle •Aluminium foil •Rubber band or electrical tape •Access to freezer.

9 Latent heat: kettle

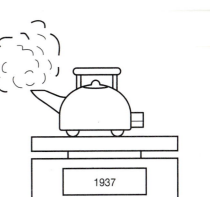

The specific latent heat of water can be measured very simply using a shiny kettle standing on a top pan balance. Fill the kettle with water, stand it on the balance and bring it to the boil. When it is boiling vigorously take the reading of the balance; continue boiling for a measured time (say 3 minutes) and record the new balance reading at the end of that time (you need a kettle without a cut-out). The difference between the two readings gives you the mass of water turned to steam in that time.

Knowing the power of your kettle enables you to calculate the specific latent heat of vaporization of water. A shiny kettle reduces heat loss and really good results have been gained by this method. A 1.5 kW kettle turns about 120 g of water into steam at 100°C.

Apparatus required: •Shiny kettle of known wattage •Stop clock •Top pan balance.

10 The freezing mixture

Really low temperatures can be obtained by adding salt to ice since the freezing point of the ice/salt mixture is lower than that of ice and so the mixture will be a liquid at the normal freezing point of water. (Temperatures of −10°C are easy to obtain). Refer to the use of salt on pavements in cold weather. The salt lowers the temperature at which the water/salt mixture freezes and so at the temperature of the air the salty water still remains a liquid.

Apparatus required: •Beaker •Ice •Salt.

11 Regelation and the ice block

Make a shoebox sized ice block. Rest it between two stools on a couple of paper towels. Hang two 1 kg masses over it on a copper wire. The wire cuts into the block but the ice then re-freezes over the top of it. This clearly demonstrates the lowering of the melting point of water when the pressure on it is increased. If the experiment is repeated with the two masses hanging on a string instead of the copper wire it won't work — the conductivity of the string is not great enough to carry away the latent heat of fusion of the ice. Refer this to the ice skater skating on water not ice, and the motion of glaciers — the meltwater between the ice and the rock.

Apparatus required: •Block of ice •Two 1kg masses •Copper wire •Tray to catch the melted water •Two lab stools as supports.

12 Boiling water under reduced pressure

Boil some water in a round flask in the neck of which is a thermometer and a glass tube fixed to a rubber tube that can be sealed with a tube clamp. When it is boiling vigorously close the clamp and turn off the bunsen immediately. Then invert the flask and pour cold water over it. Water condenses in the flask thus reducing the pressure and boiling recommences. Further cooling gives a further reduction in pressure and boiling can be obtained down to 40°C. I have even once had water boiling in a flask in the lab at 'body temperature' (37°C). (Safety screen advised here.)

Dependence of the saturated vapour pressure of water on its temperature: 37°C: 0.06×10^5 Pa; 60°C: 0.19×10^5 Pa; 75°C: 0.38×10^5 Pa; 85°C: 0.57×10^5 Pa; 100°C: 10^5 Pa.

Apparatus required: •Round-bottomed flask with bung, tube and thermometer fitted •Water •Bunsen •Retort stand and clamp •Safety screen •Tray.

13 Change of volume of melting ice: cast iron flask

A cast iron flask is filled with cold water and the lid screwed on tightly. The flask should then be put in a plastic beaker filled with a freezing mixture (alternate layers of crushed ice and salt) (see Specific/Latent Heat, experiment 10). After a while (some minutes) a crack should be heard. The flask has broken due to the increase in volume of the water as it freezes, hence the need for a plastic beaker. Relate this to the problem of burst pipes in winter.

This is also worth doing using a small plastic bottle which has been filled to the brim with water. It could be left in the freezer from one lesson to the next if you can't get one that will fit in the freezing mixture in a beaker.

Apparatus required: •Plastic beaker •Cast iron flask •Freezing mixture.

14 Change of volume of melting ice: burette

Fill a conical flask with melting ice. Put a bung in the top with a burette in it. The burette should be partly filled with light oil. As all the ice melts and a decrease in volume occurs the level of the oil in the burette should fall. Alternatively the flask could initially be filled with pure water and then placed in a freezing mixture and the resulting expansion measured as the water freezes.

Apparatus required: •Burette •Oil •Freezing mixture •Conical flask •Bung.

15 Boiling under reduced pressure: an alternative method

A simpler method of demonstrating the boiling of water at low temperature is to draw some water at a temperature of about 50–60 °C into a syringe so that about 20% of the volume of the syringe is filled. Then when the end of the syringe is sealed and the syringe is expanded rapidly the water boils under reduced pressure.

Apparatus required: •Syringe •Hot water.

16 Fish in a pond

Diagrams of the fate of fish due to the freezing of ice emphasize the importance of the relative densities of ice and water. Since ice is less dense than water at 0°C the ice floats; the water with the greater density — that at 4°C — will sink to the bottom. This is just as well, otherwise the oceans would freeze up from the bottom upwards, seriously reducing the amount of seawater!

17 Floating ice cube

Float a pure water ice cube in a beaker of water with a paper scale fixed to the outside. Allow the ice to melt and actually see if the water level goes down when the ice melts. There should be no change in level; the ice displaces its own weight of water and so when it melts the water it forms should take up exactly the same volume as the water it displaced. Try it in brine (salty water); what would you expect to happen now? What about the change in sea levels when the polar ice caps melt? Is there a difference between the effect of the Arctic and the Antarctic?

Apparatus required: •Ice block •Beaker of water •Ruler •TV camera if possible using a parallel sided plastic container for the water.

THERMAL EFFECTS AND MOLECULES

General theory for this section

The temperature of a gas depends on the kinetic energy of its molecules. $PV = \frac{1}{3}mNc_{rms}^2 = nRT$; kinetic energy of a molecule = $[\frac{3}{2}]kT$, where R is the molar gas constant = 8.3 mol^{-1} K^{-1} and k is Boltzmann's constant = 1.38×10^{-23} J K^{-1}.

1. Brownian motion and the luminous elephant
2. Porous pot and diffusion
3. Random walk
4. Diffusion in gases
5. Silent kinetic theory
6. A simulated damp proof course
7. Mixing alcohol and water

---

1   Brownian motion and the luminous elephant

The smoke cell experiment where particles of smoke are seen to exhibit a juddering motion when viewed under a microscope is an excellent way of showing the random motion of air molecules. The kinetic theory model containing ball bearings in a cylinder that can be vibrated is a useful simulation of this behaviour. Increased vibration simulates an increase in temperature as the ball bearings move faster. There are also various analogies that can be used to demonstrate this effect.

Imagine an elephant (representing a smoke particle) painted with luminous paint in a darkened sports hall. Also in the hall are a large number of children (representing the air molecule), dressed in black so that they are invisible when an observer looks into the sports hall through a skylight in the roof. The children are now told to run around, they do so, colliding with each other, the walls and the luminous elephant. What does the observer see? Simply a luminous elephant juddering about as it is being knocked from side to side by some invisible force.

---

2   Porous pot and diffusion

The different rates of diffusion of gas and air can be shown by using a porous pot. Put a bung into the top of the pot with a tube in it and fill the pot with gas from the gas tap. Upturn the pot so that the tube goes into water in a beaker. The gas diffuses out through the pot quicker than the heavier air diffuses in and water rises up the tube. Alternatively hold a glass beaker over the porous pot and fill the beaker with gas. Bubbles of gas now come out through the water.

Theory: Graham's law states that the rate of diffusion of a gas is inversely proportional to the square root of its density. Heavier gases will diffuse more slowly than light ones.

Apparatus required: •Porous pot •Beaker •Bung and tube •Beaker of water •Retort stand and clamp •Gas supply.

3 Random walk

The random nature of gas molecule motion can be shown by using isometric graph paper and a dice. The six-sided dice represents the six possible directions of motion in a three-dimensional world. The isometric paper has six possible directions from any intersection on its surface and can be used to represent a random walk, rather like the motion of a drunk in a crowd!

4 Diffusion in gases

Two and a half possible demonstrations.
   (a) Open a bottle of perfume at the front of the lab and ask the class to record when they begin to smell it!
   (b) There is also the classic experiment with concentrated ammonia and hydrochloric acid placed on plugs of cotton wool at opposite ends of a glass tube. After a while a white ring forms in the tube due to the interaction of the two chemicals. The differing masses of the molecules means that they diffuse through the air in the tube at different rates and the white ring formed in a tube when they meet will not be in the centre.
   (c) Simply opening the two bottles close to each other will show the diffusion, 'smoke' coming from the ammonia after a while

Safety: take every precaution when handling the bottles of concentrated ammonia and hydrochloric acid.

Apparatus required: •Bottle of perfume •Concentrated solutions of ammonia and hydrochloric acid •Metre long glass tube •Cotton wool •Retort stand and clamp •Dropping pipettes •Two protective saucers.

5 Silent kinetic theory

The random motion of particles in a gas can be simulated simply by sprinkling some camphor particles onto the surface of some water in a glass dish resting on an overhead projector. The camphor dissolves irregularly and the result is that the particles rush around randomly, just like the molecules in a gas. As more dissolves the motion slows down, giving a fair simulation of a gas cooling.

Apparatus required: •Overhead projector •Camphor crystals •Large dish of water.

---

6 A simulated damp proof course

Many of us know how important a damp proof course is in preventing water entering our houses. You can make a simple simulation of this by building two walls of sugar cubes in a tray with a piece of plastic in one wall between the first and second layers of cubes. Now pour some coloured water into the tray and watch the diffusion of coloured liquid up the two walls and the effect of the plastic layer 'damp course'.

Apparatus required: •Sugar cubes •Coloured water •Tray.

---

7 Mixing alcohol and water

This simple experiment gives an idea of the gaps between molecules. Take 100 ml of water in a 250 ml measuring cylinder and add exactly 100 ml of alcohol to it. The resulting volume is only about 195 ml — the molecules have intermingled, filling up the gaps between each other. Adding sugar to water is another example.

Apparatus required: •200 ml measuring cylinder •100 ml measuring cylinder •Alcohol •Water.

MISCELLANEOUS HEAT

1. Galileo thermometer
2. Boiling water in a paper bag
3. Heat sensitive mat – liquid crystals
4. Effect of heat on a rubber band
5. Clouds in a bell jar
6. Red noses – thermal effects
7. Stretching a cooled elastic band
8. Silt meter
9. Polystyrene
10. Entropy increase
11. Absolute zero
12. Dipping bird
13. Mechanical energy: heat energy conversion
14. Hot air balloon: candle
15. A good huddle
16. Floating bubbles

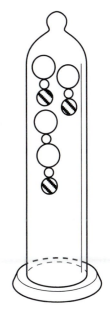

1 Galileo thermometer

This is a rather expensive (some £40 upwards) but beautiful use and demonstration of the effect of temperature on liquid density and flotation. A series of glass spheres are immersed in a liquid of density very similar to that of the spheres. They each contain different amounts of air and have small metal tags of differing mass hanging from the bottom and so rise and fall as the temperature changes. More spheres are at the top of the cylinder when it is cold and the density of the liquid is high and gives more upthrust. As the liquid heats up so the upthrust on the spheres decreases and they begin to sink to the bottom.

2 Boiling water in a paper bag

Water can be boiled in a paper bag, but why doesn't the paper burn? This is also easy to demonstrate using a paper cup — the water in the cup keeps the temperature of the paper below that required for ignition. (See *Fahrenheit 451*, a science fiction film of the 1950s.)

Apparatus required: •Bunsen burner •Paper cup and paper bag.

3 Heat sensitive mat: liquid crystals

This mat with a layer of liquid crystals below a transparent top is available from gift shops. Some lovely coloured effects are produced when a mug of hot drink is put on it or when you just lay your hand on it. Some idea of the temperature of various students' hands can be gained by the colours that you get when they put their hands on the mat.

4 Effect of heat on a rubber band

A long rubber band with a mass on the end is hung in a tall beaker of water. The taller the beaker, the longer the rubber band or piece of elastic and the more of it immersed in the water the better. The water is then heated and the change in length of the band (it shrinks) is measured. A way of getting the longest piece of rubber under the water is to use a large glass tube with a rubber bung in the lower end heated in a water bath, the water bath being a tall one-litre beaker. The temperature of the water near the rubber band is found by hanging a thermometer down the large glass tube.

Apparatus required: •Long thin rubber band •Glass tube (diameter 5 cm, length 0.5 m) •Tall beaker •Bung •Bunsen burner •Tripod •Heat-resistant mat •Thermometer.

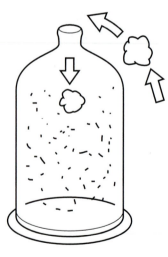

5 Clouds in a bell jar

Clouds can be formed in a bell jar if the air inside it is allowed to expand adiabatically. Take a bell jar and put 1 cm of water in the bottom and then blow into the bell jar. The air inside is compressed and as it expands you might expect to see clouds due to the adiabatic expansion and subsequent cooling, but none form. To get the cloud you need a little pollution to give nuclei on which the water can condense. This can be provided in the form of smoke. If a little is puffed into the jar clouds immediately form. (You can warm the jar to encourage evaporation).

   Another way of getting the clouds after blowing into a flask is to drop a smouldering match in to give the nuclei and thence produce a cloud.

Apparatus required: •Bell jar •Rubber tube •Smoke generator.

6 Red noses: thermal effects

The plastic red noses from various years of Comic Relief Day in England show wonderful thermal effects, the plastic changing from red to yellow or red to pink when they are put in hot water. You can also get various plastic animals that will do the same. I have some model dinosaurs that change from brown to green when heated and even a couple of small plastic dogs that have a red 'poorly' paw when put in a beaker of icy water. Rubbing the paw with your finger to 'Make it better' warms the paw and the red mark disappears!

7 Stretching a cooled elastic band

The effect of a change of temperature of a rubber band can also be investigated using a cooling experiment. Hang up a rubber band and then stretch it using slotted masses suspended from the lower end. Cool sections of the band using a freezer spray and observe the widths of the warm and cool sections. I have been told that when your lips are compressed in a kiss they actually cool but I leave it to your ingenuity to devise an experiment to show this!

Apparatus required: •Rubber band •Slotted masses •Freezer spray •Retort stand and clamp.

8 Silt meter

This experiment is a simulation of river pollution. Each group has a large beaker of water to which they add drops of milk. They place a low-voltage bulb one side of the beaker and observe the transmission of light through the liquid, detecting the intensity with a light dependent resistor (LDR). As more milk is added, and it doesn't need much, the intensity of the transmitted light falls, simulating the effect of a polluted river. The milk is less messy than mud. This has been found to be a good assessed practical investigation and could be extended by asking the pupils to estimate the concentration of milk in an unknown sample using their calibration graph from the first part of the experiment.

Apparatus required: •Light dependent resistor and ohmmeter •Calibration graph •Beaker of water •Milk and dropper •Power supply •Bulb.

9 Polystyrene

This material is yet another example of the effect of a poor thermal conductor like a carpet. If you walk on a carpet in bare feet it feels warm, but if you step off onto a tiled floor it feels much colder although the temperature of the floor and carpet are the same. Similarly a piece of polystyrene feels warm to the touch, although it is at the same temperature as other objects in the lab. However, a large piece of metal initially at the same temperature as the polystyrene will feel colder to the touch.

Both these facts can be explained by realizing that the tiles and the metal conduct heat energy away from your body leaving your hands or feet cooler. A quantitative experiment to demonstrate this can be done by placing identical beakers of water on a slab of polystyrene and metal and measuring their rate of cooling.

Apparatus required: •Slab of polystyrene foam •Large slab of metal the same size as the polystyrene if possible •Thermometers •Beakers •Stopclock.

10 Entropy increase

Entropy or disorder must always increase! This can be shown by running a video of a pile of falling blocks backwards. Since entropy always increases you can tell if the film is running forwards or not by the increase of disorder that results if no external human influences act. You can refer this to the ultimate fate of the universe – a smeared out warmth is the likely final state rather than discrete hot spots.

11 Absolute zero

You may know that one of the laws of thermodynamics is that absolute zero ($-273.15°$C) can never be reached. In fact, it gets more and more difficult to remove energy as the temperature gets lower — the energy steps become smaller and smaller. It is rather like going down a never-ending escalator; the steps on the escalator also get smaller and smaller towards the bottom and in the temperature case the energy steps go on doing this for ever.

12 Dipping bird

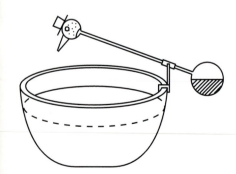

This wonderful old toy can still be bought. It rests on the side of a glass and dips its head into the cold water. As the liquid inside its body evaporates it forces liquid up the neck altering the balance of the bird which then dips its head into the water.

If the head is not wetted then a very instructive version of the experiment occurs. The bird tends not to drink but placing a saucer of hot water beneath its bottom will heat up the liquid sufficiently increase the rate of evaporation and make it start rocking. Energy is provided by the hot water — you don't get something for nothing!

Try it using alcohol instead of water. The evaporation is greater and the bird dips more quickly.

Apparatus required: •Dipping bird •Wine glass.

13 Mechanical energy: heat energy conversion

A couple of suggestions to show the conversion of mechanical energy to heat energy are:

(a) saw through an insulated bolt and let the pieces fall into a beaker of water; of course the hacksaw gets hot as well!

(b) hammer a piece of lead and then measure the temperature rise with a thermistor.

14 Hot air balloon: candle

Try making a hot air balloon from a large plastic bag and then 'flying' it by putting it over a candle; there is rather less heat energy than using a bunsen burner and the experiment is therefore easier to control.

15 A good huddle

To show the effect of groups of animals keeping warm prepare seven or eight test tubes, all containing the same amount of water. Place them in a water bath (this could just be a large beaker of water), heat them to the same temperature and then remove them – leaving one on its own and the others in a group. Record the temperature of each test tube with time; the one on its own will cool down much quicker than those in the huddle since it has a much greater surface area to volume ratio.

An interesting extension of this experiment is to investigate the effect of mass and surface area on the rate of cooling using balloons filled with hot water. Measure their temperature with a thermometer probe placed inside them and experiment with:

(a) different-sized spherical balloons,

(b) spherical balloons and sausage-shaped balloons of the same mass,

(c) spherical balloons of the same size and mass but one suspended in air and the other suspended in a bucket of water at room temperature.

Apparatus required: •Test tubes with little or no flange at the top •Water bath •Thermometers •Suitable method of holding a group of test tubes •Various balloons •Bucket •Thermometer probe.

16 Floating bubbles

The following experiment is a fascinating demonstration of the insulating properties of a gas (in this case steam). Place a metal plate about 10 cm in diameter and 0.5 cm thick on a gas burner (better than a bunsen as the whole plate gets heated). Heat the plate strongly and then drop a small droplet of water onto it. The water evaporates immediately. However, when the plate reaches about 200°C the water evaporates so rapidly from the lower surface that the resulting steam insulates the upper part of the droplet and it does not evaporate. When this state is reached more water can be added so increasing the size of the droplet which now 'floats' on the insulating layer of steam. Discs of water up to 2 cm in diameter can be produced.

Apparatus required: •Gas burner •Dropper •Metal plate •TV camera if available.

CURRENT ELECTRICITY

General theory for this section

Resistance = voltage/current;

resistance = resistivity × length/cross-sectional area.

For a metal, resistance = resistance at 0°C [1 + temperature coefficient of resistance × temperature change].

Average effects of continuous ac or dc electrical currents on healthy adults

Electrical current	Biological effect
1 mA	threshold for feeling
10–20 mA	voluntary let-go of circuit impossible
25 mA	onset of muscular contractions
50–200 mA	ventricular fibrillation or cardiac arrest

The figures given above also depend on the path of the current through the body. For example if you are outside in a thunderstorm it is important to give a possible path for a lightning strike between the highest part of your body and the ground that would not pass through your brain or your heart.

1. The electron steeplechase
2. Fusing currents in wires
3. Surge of current in a light bulb
4. Hot wire ammeter
5. Conductivity of glass
6. Lie detector
7. Resistance: conducting putty
8. The human battery
9. Repulsion of aluminium strips
10. Parallel circuits
11. Birds on high voltage wires
12. Flow of charge
13. The lemon battery
14. Parallel circuit: the bath analogy
15. Bulbs in series
16. Making ac and dc visible
17. A carbon resistor and heat: semiconductor or not?

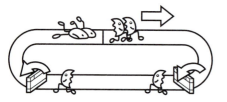

1   The electron steeplechase

This analogy is designed to explain energy losses in a series circuit. Imagine that the electrons are running round a steeplechase course. As they go round they lose energy; on the track it would be by going over a barrier, in a circuit it might be by passing through a bulb. When they reach the end of the track, or circuit, they have low energy and must be given a further input of energy — say by a battery — before they can make another circuit. Two important rules about them: (i) they are not allowed to give up the race; the same number of electrons that leave the start must reach the finish (this emphazises the constancy of current at all points in a series circuit) and (ii) the velocity of the electrons remains unaltered. We assume that they do not lose much, if any, energy on the flat — this is analogous to the low energy loss in the connecting wires of a circuit due to their low resistance.

2   Fusing currents in wires

Find the current at which a wire will fuse. Either by a steady increase or by the sudden application of a large current. Wire wool works quite well for this although a piece of resistance wire stretched between two retort stands glowing bright yellow before it fuses is quite spectacular. For older students a consideration of the radiation emitted by the wire will lead to a study of Stefan's law (energy emitted by a 'black body' at absolute temperature $T = \sigma A T^4$ where A is the surface area and σ is Stefan's constant). Factors such as the surface area of the wire and the ratio of fusing current to wire diameter could be investigated.

Apparatus required: •Wire wool •Resistance wire •Ammeter •12 V dc power supply (8 A) •Heat-resistant mat.

3   Surge of current in a light bulb

(a) Use a computer sensor to investigate the surge of current when a light bulb is switched on. Mention that switching on is the time when it is most likely that the bulb will blow. The filament has a low resistance when cold and therefore a large current flows, heating occurs rapidly, there is a large thermal expansion and the resulting thermal shock can break the filament.

   (b) The change of resistance of the filament of a bulb with temperature can be seen very easily as an extension to experiment 3(a). If the supply is adjusted to around 100 V and then switched on the small bulb lights brilliantly. The current is large since the 60 W bulb is still cold and its resistance is low. As it warms up (this takes a little less than a second) the resistance rises and the current falls and the small bulb's brightness decreases.

Theory: resistance of a metal conductor increases with increasing temperature. As the temperature rises the thermal motion of the atoms within the metal impedes the motion of electrons through it and so the resistance rises.

Apparatus required: •Mains bulb in suitable holder •Computer •Digital voltage and current sensor with optional link to a computer.

4   Hot wire ammeter

Suspend a taut piece of copper wire between two retort stands with its ends fixed to insulated terminals. Hang a 50 g mass from the centre to keep it in tension. Pass a current through it and measure the depression of the mass. Plot a graph of depression against current. Note that this will work for both ac and dc.

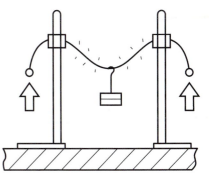

Theory: As a current passes through the wire it will heat up and thus expand. The weight hanging from its centre will cause it to sag.

Apparatus required: •Copper wire 28 SWG •Retort stands and clamps •12 V dc supply giving up to 8 A •50 g mass on metal hanger •30 cm ruler in bench clamp •TV camera if possible.

5   Conductivity of glass

This is a fascinating demonstration. Take two pieces of thick copper wire and wind one round each end of a soft glass rod; connect the ends of the wires to a 240 V mains power supply in series with a 100 W lamp. Switch on — nothing happens since the glass is an electrical insulator. Now heat the glass strongly with a bunsen. As the glass becomes molten conduction occurs and the light comes on. It is much more impressive with the mains where once conduction has started the very heat of the current itself is often enough to maintain the current flow.

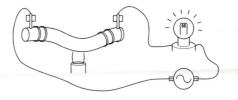

   This experiment has some important safety considerations since mains electricity is used, in part unshielded. A low-voltage version using 12 V can be attempted but it is nowhere near as impressive.

Apparatus required: •Variac transformer •Soft glass rod •Bunsen burner •Two lengths of bare copper wire •Wooden clamp •Crocodile clips •Heat-resistant mat •100 W lamp in holder •Three insulated in-line connectors.

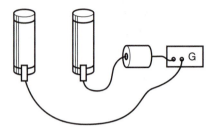

6 Lie detector

This experiment is a fun extension of resistance and simulates a lie detector or embarrassment meter! Wrap pieces of aluminium foil round two toilet rolls or tape two sheets of foil to the bench. Connect a 1.5 V cell and a spot galvanometer in series with the foil and then complete the circuit by getting one of the pupils to hold the toilet rolls. Now try and embarrass them; the conductivity of the skin changes, and as a result the reading on the spot galvanometer changes also. Quite often the volunteer can be embarrassed just by sitting there — they know what secrets about them are known by the rest of the class. Resistance between the two hands is of the order of 200 kΩ.

Be sensitive here — I never let anyone's private life get exposed to the others. The same effects occur if the volunteer is put under stress — try asking them some physics questions!

(Warn about never doing this outside school; only use the 1.5 V and never connect it near to a student's head.)

Apparatus required: •Two cardboard tubes •Electrical tape •Aluminium foil •Spot galvanometer •1.5 V cell •Crocodile clips.

7 Resistance: conducting putty

Use this commercially available material to make different shaped specimens to demonstrate the effect of shape and size on the resistance of a specimen. Even series and parallel circuits can be moulded. Use a couple of metal discs pressed onto either end of the specimen as a means of making electrical contact.

Apparatus required: •Conducting putty •12 V dc power supply •Ammeter and voltmeter •Crocodile clips •Two metal discs.

8 The human battery

Place one hand on a copper sheet and the other hand on a zinc sheet and measure the potential difference between them. Voltages of about 0.7 V can be produced due to the electrochemical reaction between the two dissimilar metals and the moisture of your hands. As with the lie detector, try the effects of stress on the 'volunteer'!

Apparatus required: •Voltmeter •Zinc sheet •Copper sheet.

9 Repulsion of aluminium strips

Two slack aluminium strips mounted vertically side by side can be used to show the forces between currents. Make sure that they cannot touch when a current is passed through them. A single folded strip will show the repulsion of currents flowing in opposite directions, down one side and up the other.

Apparatus required: •Two aluminium strips •Holders •Power supply •Ammeter.

10 Parallel circuits

Pupils are taught that the addition of a resistor to a circuit will decrease the flow of current and are therefore puzzled when told that two resistors in parallel draw a greater current from the supply than one of the resistors on its own. You can use the traffic analogy to explain this. The addition of a bypass round a town will actually increase the traffic flow on the major road. Similarly, another resistor in parallel with the first will allow more current to flow from the supply. (See also the bath plug analogy (experiment 14).)

11 Birds on high voltage wires

Why is it that birds can sit on high voltage cables without danger? Of course it is fine if they do not touch the ground or have one foot on the live cable and one on the neutral. They are quite safe because the potential difference between their two feet when standing on a single cable is far too low to be dangerous.

12 Flow of charge

This experiment is designed to measure the drift velocity of ions in a liquid and gives an idea of the drift velocity of electrons in a piece of metal when a potential difference is placed across its ends. Soak a rectangle of filter paper in ammonium hydroxide solution and lay it on a microscope slide. Place an optical pin across each end and clip them in place with crocodile clips.

   Connect an HT supply across them and place a potassium permanganate crystal in the centre of the paper. Apply a potential difference of 100 V between the crocodile clips and measure the time it

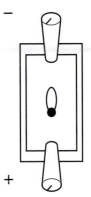

takes for the purple colour to travel 1 cm. Doing the whole experiment on a plastic strip with a half millimetre scale marked on it helps and using a TV camera to view the whole slide on a screen helps even more.

Theory: in a wire the current is given by the equation $I = nAve$ where v is the drift velocity of the electrons, A the cross sectional area of the wire, n the number of electrons per metre cubed and e the electron charge. Similar values to v can be obtained by this drift velocity experiment. Values of a few mm per minute are common for potential differences of 100 V between electrodes some three or four cm apart.

Apparatus required: •HT power supply (50–200 V) •Crocodile clips •Two optical pins •Microscope slide •Filter paper •Millimetre scale •Stop clock •Potassium permanganate crystals •Ammonium hydroxide.

13 The lemon battery

Put two dissimilar metal electrodes (for example copper and zinc) into a lemon. A potential difference of about 1 V should be obtained between them (lemons usually give about 1.08 V and oranges 0.95 V). This can either be detected by using a low voltage bulb (more impressive) or a digital voltmeter. I have not really been successful recently with lighting a bulb but the voltmeter did show that there was something happening.

A way of making this experiment more impressive is to use a set of lemons in series to trickle charge a capacitor. Then discharge this through an LED to give a flash of light like the camera flash.

Apparatus required: •Lemon •Electrodes of copper and zinc •Crocodile clips •Low-voltage bulb •LED RS 586 447 •Capacitor 220 mF •25 V dc voltmeter.

14 Parallel circuit: the bath analogy

Imagine a bath full of water but with two plugs and plug holes. The plughole allows water to fall into a tank where a pump pumps it back to the tap. Now turn on the tap and pump and pull out one of the plugs. Water circulates round and the level of the water in the bath stays the same.

Now pull out the other plug. To keep the water level in the bath the same the pump must work twice as hard — the rate of flow of water from the tap is doubled but the rate of flow of water from the first plug hole is unaltered. This is a good analogy with a parallel circuit, the two plugholes representing the branches of the circuit, the pump replacing the battery and the tap the wires before they branch. The height of the water in the bath is analogous to the potential in the circuit.

15 Bulbs in series

I always introduce the idea of voltage by comparing the drop in potential round a circuit containing a number of bulbs with the drop in potential energy down a flight of stairs. Each stair representing one of the bulbs. From there I go on to show two experiments to demonstrate this.

In the first one connect a 2.5 V torch bulb in series with a 12 V, 24 W bulb. Steadily increase the potential difference across the circuit and show that the small bulb does not blow. They both carry the same current but the greatest potential difference drop occurs across the large bulb. It is much more impressive if this can be done with a 2.5 V, 0.6 W (0.25 A) torch bulb and a 240 V, 60 W (0.25 A) mains light bulb, the current being about 0.25 A in both. Safety precautions are necessary here to ensure that nobody touches the bare mains terminals.

In the second experiment twenty 2.5 V torch bulbs are connected in series to a 50 V, 1 A supply. They all light — the same current passes through each — and the potential difference drop across each is roughly the same, about 2.5 V.

Apparatus required: •Mains bulb 60 W, 0.25 A •Twenty torch bulbs 2.5 V 0.25 A •Variac •Ammeter 0–10 A (digital).

16 Making ac and dc visible

Take a piece of thick blotting paper and soak it with some thinly running paste of starch (or flour) mixed with potassium iodide. Allow it to dry and then mount it on a metal base. Connect the base to one terminal of the supply using a crocodile clip. Run a probe connected to the other terminal of the supply along the paper surface. First use a dc supply and then an ac supply. The dc supply will give a continuous line while the ac will give a series of dashes.

The action of the electricity releases iodine ions which move towards the positive terminal where they react with the starch, giving a dark blue colour.

Apparatus required: •Blotting paper •Starch or flour •Probe •Power supply •Potassium iodide crystals •Crocodile clips •Metal sheet.

17 A carbon resistor and heat: semiconductor or not?

The effect of raising the temperature of a carbon resistor can be investigated in the following way. Take a 10 MΩ resistor, connect it to a high resistance ohmmeter and mount it in a holder. Use a hairdryer to heat the resistor and measure the resistance as the temperature changes.

The resistance is first seen to fall and then rise again.

Theory: at first the semiconductor nature of carbon is more important —
more free electrons are created and the resistance falls. As the
temperature gets greater the increased thermal motion gives a greater rise
in the resistance; a minimum of resistance is reached and then it starts to
rise.

Apparatus required: •Hairdryer •10 MΩ resistor •Mount •High resistance
ohmmeter.

MAGNETISM

General theory for this section
Like poles repel and unlike poles attract. The force decreases with increasing separation.
Common magnetic materials: iron, steel, cobalt, nickel.

1. Magnetic fields and tape recorders
2. Magnets and a top pan balance: Newton's third law
3. Magnetic field patterns
4. Glass rods to hold floating magnets
5. Three dimensional magnetic fields
6. Executive magnetic toys
7. Curie effect: iron
8. Indian rope trick
9. Magnetic force

1 Magnetic fields and tape recorders

(a) Sound is stored on a magnetic tape as a pattern of tiny magnetic domains within the tape. This can be demonstrated clearly by showing the effect of a magnet on a recorded tape. Record some speech on a cassette tape and then pass the cassette between the poles of a strong magnet. Now replay the tape; the magnetic pattern will have been altered and the sound will be badly distorted or even completely erased. The tape can be used again afterwards. If you have a video recorder try this using videotape — some interesting effects on the picture when it is replayed will result.

(b) For the second part of the experiment you will need a reel-to-reel tape recorder, some iron filings and a signal generator. Use the signal generator to record a 50 Hz sine wave signal on tape. Remove that section of tape from the tape recorder and sprinkle some fine iron filings on it. The filings will adhere to the tape, more strongly in places where the signal was strongest, i.e. at the troughs and peaks of the sine wave. Different wave shapes and different frequencies may be studied. Fast tape speeds will show the effects much more clearly, of course.

(c) A simulation of the effects of tape recording using magnetic tape can be achieved using adhesive tape coated with fine iron powder.

Apparatus required: •Reel-to-reel tape recorder •Microphone •Strong magnet •Blank tape •Iron filings •Signal generator or musical tube of variable pitch.

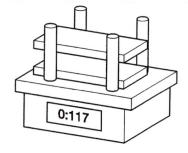

2   Magnets and a top pan balance: Newton's third law

The forces between two magnets can easily be shown by mounting a magnet on a top pan balance and lowering another towards it, guided by glass rods. This is also a good example of Newton's third law.

Apparatus required: •Magnets •Suitable holder such as glass rods mounted vertically in a wooden block •Top pan balance.

3   Magnetic field patterns

The magnetic field of a permanent magnet can be demonstrated by one of the following experiments.

(a) Put a bar magnet on the glass of an overhead projector and cover it with a piece of acetate film. Sprinkle some iron filings on the film and project the image onto a screen for class observation. The acetate sheet prevents the filings clinging to the magnet. Coils of wire mounted in plastic holders are commercially available to demonstrate electromagnetic fields.

(b) Another way of making a permanent record of the magnetic field of a magnet is to do a similar experiment on photographic paper in the dark room with just the safe light on. Put a piece of photographic paper over the magnet and, after sprinkling the iron filings on the paper to get a good field pattern, turn the main light on for a few seconds. Then develop and print the paper and use the silhouette photos of the magnetic field patterns to retain the image.

(c) This experiment is another way of producing a permanent record of the magnetic field of a magnet. Prepare some sheets of waxed paper by dipping them in molten wax. This is easy to do by having a tin of molten wax heated over a bunsen. Lay the magnet on the bench and place a piece of waxed paper on it. Sprinkle some iron filings onto the paper until a good field pattern is seen. Now for the difficult part. Carefully lift the paper vertically off the magnet, keeping the paper horizontal and melt the wax gently by passing it smoothly but quickly backwards and forwards through a clear blue bunsen flame. Remove it from the flame and allow the wax to solidify so giving a permanent field pattern. Finally fix it into a book using clear book covering film to preserve the pattern.

With all three of these experiments tapping the edge of the paper slightly helps to move the iron filings and so improve the pattern.

Apparatus required:
(a) •Magnet •Iron filings •Overhead projector •Acetate sheet.
(b) •Magnet •Iron filings •Photographic paper •Access to a dark room •Photographic chemicals.
(c) •Magnets •Iron filings •Waxed paper •Bunsen.

4 Glass rods to hold floating magnets

Glass rods mounted vertically in a wooden base are a good way of holding permanent magnets for magnetic levitation experiments. They are non-magnetic and there is little friction between them and the magnet.

5 Three-dimensional magnetic fields

The three-dimensional magnetic field of a magnet can be investigated by using a jelly. Make up the jelly in a clear mould and then add the iron filings before it sets. Stir well and then apply the field, or pour it into a mould which has a magnet in it and allow the jelly to set. (Various recipes available!) I am not really sure how long they keep without going mouldy! Maybe somebody could suggest an alternative material.

Apparatus required: •Jelly •Magnet •Clear mould •Coil and power supply •Iron filings.

6 Executive magnetic toys

These are just good to look at but every school should have one or two. They stimulate much discussion about the magnetic effects, the balancing of the parts and the low friction at the points of contact.

Apparatus required: •Executive magnetic toys.

7 Curie effect: iron

The effect of heat on the ability of a bar to become magnetized can be studied using the Curie effect. Place a rod of iron on a tripod with a small magnet hanging from its end. Heat the rod strongly with a bunsen flame; when the rod is hot enough the magnet will fall off. If you can spare the magnet it is even better to heat the magnet itself. The motion of the magnetic domains is sufficient to destroy the overall magnetism of the magnet and prevent the rod being magnetized.

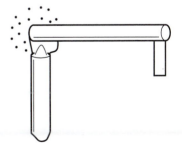

Apparatus required: •Iron bar •Magnet •Bunsen •Tripod •Heat-resistant mat.

8 Indian rope trick

Have a magnet concealed in a tube which is held in a clamp above a small steel screw that is fixed to the bench by a thread. The length of the thread should be a little more (a few centimetres) than the height of the magnet above the bench. The screw can then seem to rise into the air without support — it is actually held up by the invisible magnet. You can pass your hand through the gap between the screw and the magnet to show no support — of course your hand does not affect the magnetic field. Then try it with a metal plate, first aluminium and then steel.

Apparatus required: •Screw •Thread •Magnet •Tube to conceal the magnet •Retort stand.

---

9 Magnetic force

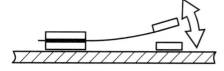

The forces between magnets can be demonstrated and indeed measured in the following way. Fix a ceramic magnet on the bench with one of its poles uppermost. Mount a hacksaw blade above this magnet, fixed down at one end with a ceramic magnet fixed to the other so that it is above the magnet on the bench with like poles facing each other; the blade will curve upwards. Measurement of the amount of curvature and a control experiment carried out by loading the blade with masses can give you an idea of the forces between the magnets.

Apparatus required: •Two ceramic magnets •Hacksaw blade •Suitable clamp •Masses •Ruler.

ELECTROMAGNETISM

General theory for this section

The strength (magnetic flux density) of an electromagnet in the form of a straight coil (solenoid) can be calculated from the formula $B = \mu NI/L$ where N is the number of coils, I is the current in amps passing through them and L is the length of the solenoid. μ is a constant depending on the material of the core. For an iron core this has a value of about 0.002 H m^{-1} while for air it is two thousand times smaller.

The strength of an electromagnet can be increased by increasing the number of coils per metre (for the same current), increasing the current flowing through it or putting a piece of soft iron in its centre.

When a wire of length L is placed at right angles to a magnetic field of strength B it experiences a force BIL at right angles to itself and the field. The bigger the field, the current or the length of the wire the bigger the force.

1. Electromagnetic forces: suspended coil
2. Strength of an electromagnet
3. Forces between currents
4. Model loudspeakers
5. Force on a current in a magnetic field
6. The catapult field
7. Faraday effect

---

1 Electromagnetic forces: suspended coil

The magnetic field produced when a current flows in a coil may be detected by hanging a light coil of wire from its connecting leads near a magnet. When a current is passed through the coil the force between the coil and the magnet will either attract or repel the coil, making it swing towards or away from the magnet depending on the direction of the current in the coil.

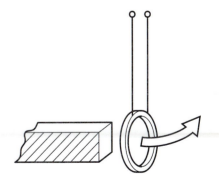

Apparatus required: •Coil of fine wire •Power supply •Magnet in non-magnetic stand.

---

2 Strength of an electromagnet

A very simple method of estimating the strength of an electromagnet is to see how many steel paper clips it will support. Connect up an electromagnet, switch it on and support a string of paper clips from the

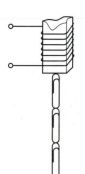

lower end. Varying the current will change the strength of the electromagnet and this can be measured by the number of paper clips that can be supported. It is important to check that the clips are not magnetized before starting the experiment!

Apparatus required: •Electromagnet •Power supply •Ammeter •Paper clips.

3 Forces between currents

A way of demonstrating the force between currents is to use a small metal slinky spring. I have bought one that has a coil diameter of only 3 cm. Hang this up vertically with the lower end free and pass a dc current through it. The currents in adjacent loops of the spring are in the same direction and so the spring contracts — like currents attract. Placing a transparent plastic ruler in front of the coils enables the change in separation to be measured as the current in the coil is changed.

Using ac can give some interesting resonance effects if the tension of the spring is varied by adding small masses to the lower end.

Theory: the force per unit length between two currents (I_1 and I_2) in two infinitely long parallel straight wires separated by a distance d is given by the formula $F = \mu_o I_1 I_2 / 2\pi d$. I know that our spring is far from being a straight wire but at least you can get an idea of the possible forces involved.

Apparatus required: •Small slinky spring diameter of coils about 3 cm •Retort stand •Fine copper wire for leads •Power supply (ac and dc) •Transparent plastic ruler •TV camera if possible •Light slotted masses (< 10 g).

4 Model loudspeakers

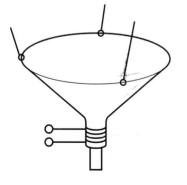

An impressive model loudspeaker can easily be made from paper, some insulated wire and four strong bar magnets. Make a large paper cone, cut a piece off the bottom and fix a short cylinder of paper to the cone in its place. Wind a coil of thin enamelled wire (say 20 turns) round the cylindrical projecting part. Place the cylindrical part over four bar magnets taped together standing vertically on the bench and with the same direction of polarity. Suspend the cone by thin elastic from three or more retort stands, connect the coil to a signal generator and switch on. Using a low frequency voltage enables the oscillatory motion of the cone to be seen clearly.

Apparatus required: •Paper •Insulated wire •Four strong bar magnets (about 8 cm long) •Power supply or signal generator •Thread •Three retort stands.

5 Force on a current in a magnetic field

The following experiments are to demonstrate the force on a current in a magnetic field.

(a) Suspend a light rod from two thin copper wires so that the rod hangs horizontally between the poles of an eclipse major magnet (a large U-shaped magnet with a flux density between the poles of about 0.5 T) and is free to swing. The field of the magnet acts vertically across the rod. Pass a current through the wires and rod and the rod should swing in and out of the field.

(b) In the second experiment a slack strip of aluminium foil is pinned to a cork mat with a large strong (0.5 T) magnet standing over it so that the strip lies between the poles. Passing a current through the strip in the correct direction makes it rise from the mat — a clear demonstration of the force on a current in a magnetic field.

(c) In the third experiment the force is measured using a top pan balance. Clamp a metal rod horizontally above a top pan balance so that it passes between the poles of a horseshoe magnet that is resting on the balance. Pass a current through the rod and use the change in the reading on the balance to demonstrate the force on the rod due to the field. Making measurements of the current in the rod and the change in reading on the balance will enable you either to calculate the strength of the magnetic field or to verify the law of force on the rod.

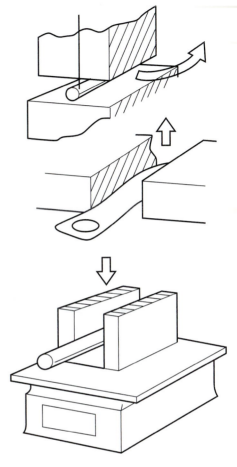

Theory: force on a current in a magnetic field = BIL where B is the magnetic flux density, I the current in the wire in amps and L the length of the conductor in the field. The strength of the large magnet that I use is about 0.5 T and that of the small magnadur magnets about 0.05 T.

Apparatus required:
(a) •Light metal rod suspended by thin cooper wires •Horseshoe magnet.
(b) •Strong horseshoe magnet •Aluminium foil strip pinned to a cork mat •Crocodile clips •Power supply.
(c) •Top pan balance •Stiff metal rod •Horseshoe magnet •Power supply.

6 The catapult field

You can compare the force on a wire in a magnetic field to that of a stretched piece of elastic or the bed of a trampoline. For example, if you jump onto a trampoline there is a resultant vertical force that catapults you upwards, at right angles to the trampoline. Since a similar motion is observed when a wire carrying a current is placed at right angles to a magnetic field the combined field produced by the wire and the magnet is often called the catapult field. You can show this easily by mounting two lengths of thick (say 24 SWG) bare copper wire between the poles of a strong U-shaped magnet so that they are horizontal and parallel. Then

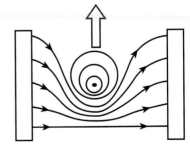

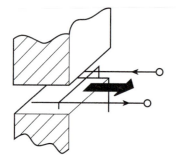

place a third wire across them so that it rests on the two straight wires. Passing a current through the wires will make the loose wire slide along; indeed if the current is big enough (a few amps) the wire will shoot off the end.

Apparatus required: •Two straight lengths (10 cm or so) of stiff clean copper wire •One short length of thinner copper wire •A high current power supply •U-shaped magnet.

7   Faraday effect

Shine a beam of light across the lab so that it passes through crossed polaroids at either end to cut off the light. Now put a strong magnet (0.5 T) across the beam and see if you can detect any change in the direction of polarization. In other words, does any light now emerge from the second polaroid?

Apparatus required: •Electromagnet •Light source.

ELECTROMAGNETIC INDUCTION

General theory for this section

When a changing current flows in a coil a varying potential difference is induced in a nearby coil. The size of the induced potential difference depends on the number of turns of both coils and their linkage — whether there is a soft iron core joining them. It is formed because of the varying magnetic field produced by the primary current and this variation can also be produce by moving a permanent magnet near a coil.

Currents are also induced in metal plates by moving a magnet near them; these are called eddy currents and the magnetic field produced by them acts so as to reduce the original motion.

1. Eddy currents
2. Electromagnetic brake
3. Induction: light bulbs and coil ac/dc
4. Magnets oscillating in a coil
5. Jumping ring and solid carbon dioxide
6. LED and coil: electromagnetic induction
7. Aluminium plate under a swinging magnet
8. Magnet in a tube: electromagnetic induction
9. Transformers: the action of a choke
10. Electromagnetic induction analogy
11. Detecting radiation
12. Thickness measurement: inductance
13. Electromagnetic separator
14. Eddy currents and the linear air track
15. Audio loop
16. Tape recorder simulation
17. Frequency of the mains
18. A light bulb in a microwave

1 Eddy currents

This experiment is a very simple way of showing eddy current damping. Suspend a copper cylinder from a thread so that it hangs between the poles of a large permanent magnet (flux density about 0.5 T). Twist the thread so that it oscillates about a vertical axis when released. The motion of the cylinder shows considerable damping because of the eddy currents set up

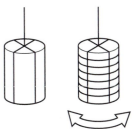

within it. Now repeat the experiment using a pile of small copper or brass coins taped together, the damping is much less because the gaps between the coins only allow much reduced eddy currents to flow in the stack.

Apparatus required: •Pile of small copper coins •Copper cylinder •Thread •Retort stand •Large U-shaped magnet.

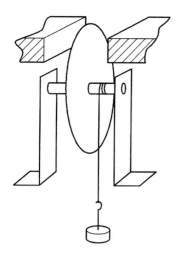

2   Electromagnetic brake

Spinning a non-magnetic metal disc between a pair of magnadur magnets mounted vertically about a cm apart with opposite poles facing each other also shows eddy current damping. The disc can be made to spin in a vertical plane by tying a piece of cotton to the axle of the disc and passing it over a pulley to a weight that is free to fall. As the weight falls it accelerates continually without the magnets but reaches a terminal velocity when they are in place.

Compare this with the air damping experiment (resonance and damping, experiment 9).

Apparatus required: •Aluminium disc on horizontal axle •Pair of magnadur magnets on steel yoke •Cotton.

3   Induction: light bulbs and coil ac/dc

The effect of self induction of a coil in a circuit can be shown as follows.

(a) Connect two bulbs in parallel to a dc supply. In one branch connect a resistor in series with the bulb and in the other connect a coil with an iron core of identical resistance to the resistor. Switch on — the bulb in series with the coil takes longer to come on demonstrating the inductive effects of the coil.

(b) Put a bulb and a coil in series with first a dc supply then an ac supply. Notice the change in the brightness of the bulb due to the inductance effects of the coil with ac.

Theory: inductance of a coil = $\mu_0\mu_r N^2 A/L$, length L, cross sectional area A and of N turns.
A coil of length 5 cm, 2 cm square, with a steel core and of 1200 turns has a self inductance of just under 30 H. With an air core this reduces to 0.15 H (150 mH).
$V = LdI/dt$ so time for the voltage across the 0.15H coil to rise to 0.2 A for a 2.5 V supply is 12 ms.
For the time to be a second or more the inductance must be 12.5 H.

Apparatus required: •Low-voltage bulbs •Ammeter, both ac and dc •Coil •Resistor.

4 Magnets oscillating in a coil

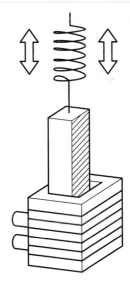

Suspend a magnet from a spring so that it hangs within a coil connected to an oscilloscope. Displace the magnet and allow it to bob up and down. The resulting induced voltage can be studied. Notice the direction of the induced voltage compared with the direction of motion of the magnet.

Ask the students whether or not they would expect to see eddy current damping in this demonstration.

Theory: since the magnet oscillates in the coil induced voltages will be produced, the size of the induced voltage being proportional to the speed of the magnet therefore giving a good demonstration of shm.

Apparatus required: •Retort stand •Magnet •Helical spring •600-turn coil •Oscilloscope.

5 Jumping ring and solid carbon dioxide

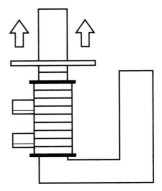

(a) The repulsion between the magnetic fields produced by two electric currents can be shown by this experiment known as the jumping ring experiment. The core of a demountable transformer is opened, a mains coil connected to the a.c mains and used as the primary and an aluminium ring as a secondary. The crosspiece is placed vertically on the arm round which the ring is slipped. Switching on the current shoots the ring into the air. As a problem for the older students, ask them what happens with a dc supply.

Theory: eddy currents induced in the ring form a magnetic field which is the same direction as that in the coil and so the ring is repelled.

The ring can be cooled by placing it in liquid nitrogen or solid carbon dioxide if the nitrogen is not available. This increases the height risen by lowering the resistance of the ring and so increasing the size of the induced current in it.

(b) Put the mains coil round the bottom of a retort stand using an aluminium ring to act as the secondary. As before you can cool the ring in solid carbon dioxide. You might get some heating problems in the retort stand due to eddy currents set up within it so maybe it is better to use a section of a laminated core as the vertical part.

Apparatus required: •Demountable transformer •Aluminium ring •Carbon dioxide cylinder •Mains coil •Mains power source •Retort stand.

6 LED and coil: electromagnetic induction

Use two light emitting diodes (LEDs) and a magnet in a coil to show the change of direction of the induced current as the magnet is moved in opposite directions through the coil. Connect the two LEDs in parallel with each other but facing opposite directions and in series with the coil. This needs a strong magnet, a large coil and low current LEDs.

Apparatus required: •Two LEDs •Magnet •Coil.

7 Aluminium plate under a swinging magnet

Suspend a bar magnet by a thread so that it hangs horizontally above an aluminium plate and start the magnet swinging. It will soon come to rest because of the induced currents in the plate. This is a simple example of eddy current damping.

Apparatus required: •Magnet •Thread •Metal plate •Retort stand.

8 Magnet in a tube: electromagnetic induction

If you have never seen this one before then be prepared for a treat. It is a most impressive example of induced currents and relies on the magnetic field produced by a falling magnet in a tube acting so as to oppose the motion of the magnet so slowing down its rate of fall. Hold a 2 m long copper tube vertically and drop a strong magnet (such as that available from Philip Harris) down the tube. It takes over five seconds compared with approximately 0.75 s when the same magnet is dropped down a plastic tube of the same dimensions.

Allowing the magnet to slide down the tube when it is inclined at an angle to the vertical is almost more impressive — it takes nearly thirty seconds to emerge depending on the angle of tilt of the tube!

(I have used the second version with the older students in a discussion of Galileo's diluted gravity experiment and the components of vectors).

Theory: Lenz's law: induced emf (ϵ) = minus the rate of change of flux in the circuit, the flux acts so as to oppose the change producing it. $\epsilon = -d\phi/dt$. Therefore the induced currents flowing round the tube produce a magnetic field along its axis which slows down the rate of fall of the magnet.

Apparatus required: •Magnet •Copper tube 2 m long if possible •Plastic tube 2 m long •Stop watch.

9 Transformers: the action of a choke

Set up a transformer with a laminated core and a step down ratio of 2:1. Put an ac ammeter in the primary circuit. With the secondary circuit open the reading on the ammeter is almost zero, it is acting as a choke; induced currents in the primary reduce the potential difference across it to almost zero. If a dc supply is used the reading is high. Putting successively more bulbs (in parallel) in the secondary circuit will increase the current in the primary. Try using a soft iron yoke to complete the circuit of the core and observe the effect on the brightness of the lamps.

Apparatus required: •Transformer – step down ratio 2:1 •AC ammeter •Bulbs •Power supply ac and dc.

10 Electromagnetic induction analogy

I use the following analogy when attempting to explain the production of an e.m.f. by the cutting of magnetic field lines. Imagine a cornfield with the combine harvester cutting the corn stalks. The cutting is more effective when the cutting edge is at right angles to the corn stalks. This is an analogy with the cutting of magnetic flux by a wire — it is more effective and so generates a greater potential when the wire is moving at right angles to the field direction.

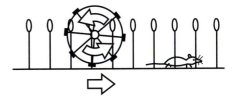

Theory: EMF generated $= BLv\sin\theta$ where θ is the angle between the wire and the field lines, L the length of the wire, v the velocity of the wire and B the magnetic flux density.

11 Detecting radiation

Use yourself as an aerial to detect mains frequency! Connect yourself to a cathode ray oscilloscope by holding a lead inserted into the Y INPUT socket. Put your other hand near to or around an insulated mains cable that has a current flowing through it. (Reaching up towards a fluorescent lamp also works.) You will see an ac trace of frequency 50 Hz on the oscilloscope screen.

Safety: on no account touch any bare wires or put your hand into a mains plug socket!

Apparatus required: •Oscilloscope.

12 Thickness measurement: inductance

The loss of energy when two parts of the core of a transformer are separated can be used as a sensitive means of measuring thickness and also as a demonstration of this energy loss. Basically it is simply an ac electromagnet with two coils separated from its steel 'keeper' by a number of sheets of paper. After calibration with paper of known thickness the voltage recorded in the secondary from a fixed primary voltage is used to measure the thickness of a sheet of paper.

Apparatus required: •Demonstration transformer with removable 'keeper' •Sheets of paper •ac voltmeter •ac power supply.

13 Electromagnetic separator

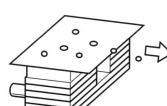

A small-scale simulation of an electromagnetic separator used to separate non-ferrous metals from other non-metallic scrap can be shown by the following experiment. Place a thin piece of card on the top of one arm of a U-shaped core of an ac electromagnet. Put a few scraps of aluminium foil on the card. When the current is turned on they will be ejected from the field because of the eddy currents within them (see the jumping ring experiment).

Apparatus required: •Electromagnet •ac power supply •Aluminium scraps (foil) •Piece of thin card.

14 Eddy currents and the linear air track

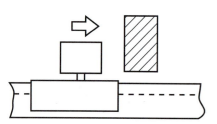

An interesting variation on the theory of eddy currents can be carried out using the linear air track. Mount a large U-shaped magnet over the track between two light gates so that a rider can pass beneath it. Mount one of the aluminium foil absorbers from the radioactivity kit on the rider so that the foil can pass between the poles of the magnet. Now accelerate the rider along the track with a constant force (by using a weight over a pulley). As the foil passes between the poles of the magnet eddy currents will be induced in it and electromagnetic braking will result. Investigate the size of the eddy currents produced for different thicknesses of foil.

Theory: since the eddy currents act to oppose the motion they reduce the acceleration of the rider. The effect of different resistances, and therefore different eddy currents, can be found by changing the foil absorbers and measuring the resulting change in the acceleration of the rider as it passes through the light gates.

Apparatus required: •Linear air track •Absorbers set •Pulley, weight and thread •Blu tac •Light gate •Large U-shaped magnet.

15 Audio loop

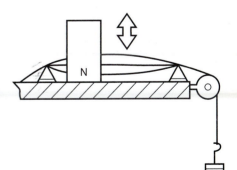

A simulation of the audio systems provided in some theatres for the hard of hearing can be produced by connecting the output of a tape recorder to a wire which is placed round the walls of the lab. The resistance of the wire loop should be made similar to that of the recommended speakers for your tape recorder. If the tape recorder is then set running the sound output signal can be detected with a second small coil connected to an amplifier placed anywhere within the lab and a pair of headphones.

Apparatus required: •Tape recorder •Coil •Headphones •Large length of wire.

16 Tape recorder simulation

Induced voltages can be shown in this simple simulation of the action of the playback head in a tape recorder. Move a 3600-turn coil over a row of ceramic magnadur magnets placed flat down on the bench with their poles alternately N–S face up with the coil connected to an amplifier and speaker or to an oscilloscope. As the coil moves a changing voltage will be induced in it and this can be detected by the speaker or oscilloscope.

Apparatus required: •3600-turn coil •At least ten magnadur magnets •Amplifier •Speaker or oscilloscope.

17 Frequency of the mains

The induced current in a wire moving in a magnetic field can be used to make a measurement of the frequency of the mains. A wire is fixed to the bench at one end, then over two glass prisms while the other end passes over a pulley and is attached to a set of slotted masses (values between 200 g and 500 g are needed for lengths of wire between about 30 cm and 50 cm). A large U-shaped magnet is placed so that the wire passes between its poles and the wire is connected to a low-voltage 50 Hz supply and a current of about 4 A is passed through the wire. Adjusting the tension of the wire can set it into resonance when its fundamental frequency equals that of the supply.

Theory: for a stretched wire the fundamental frequency f is given by the equation $f = \frac{1}{2}L(T/m)^{1/2}$ where T is the tension in the wire and m is its mass per unit length.

Apparatus required: •Large magnet (field strength = 0.2 T) •Wire •G-clamp •Bench pulley •Set of 100 g slotted masses •Ruler •Two glass prisms •ac power supply.

18 A light bulb in a microwave

Put a 100 W light bulb in a microwave oven and turn the oven on for a couple of seconds (no more!). The filament glows brightly and then breaks. It is acting as an aerial, a potential difference being induced in it due to the microwave radiation. Very impressive, but I understand that the bulb explodes if left in for too long! What would happen with a small heater filament from an electric fire?

Apparatus required: •Microwave oven •100 W light bulb (higher wattages are even better).

ELECTROSTATIC PHENOMENA

General theory for this section

Radial electric field: field intensity E at a distance r from a charge $E = \frac{1}{4\pi\epsilon_0}[Q/r^2]$.

Uniform field (i.e that between two parallel plates, separation d and potential difference V): field intensity $= V/d$.

Force between two charges (Q_1 and Q_2) $= \frac{1}{4\pi\epsilon_0}[Q_1Q_2/r^2]$.

Force on a charge Q in a uniform field $= QE = QV/d$.

1. Electric fields
2. Induction and repulsion in electrostatics
3. Plastic box smoke separator
4. Drawing pin on the van de Graaff: dust/smoke collector
5. Some simple electrostatic demonstrations
6. The charge on a body
7. Bubbles and the van de Graaff
8. A charged rod and its effect on water streams
9. Electric fields
10. Spark plug and the van de Graaff
11. Electric fields: the school dinner experiment
12. Spark counter and ions from a flame
13. Electrostatic fields
14. Watch glass electrostatic repulsion
15. Electrostatic forces: balls and a top pan balance
16. van de Graaff and a radio
17. Soap bubbles and Millikan
18. Electrostatic repulsion: various versions
19. Fluorescent tube and the plasma globe
20. Statics and covering film
21. Electrostatics and the photocopier
22. Ions in a flame and a flame probe for detection of electrostatic fields
23. Point discharge
24. Pith ball on thread and the van de Graaff
25. Paper and the van de Graaff
26. The phantom leg
27. Field in a hollow charged conductor
28. Cars and electrostatics

1   Electric fields

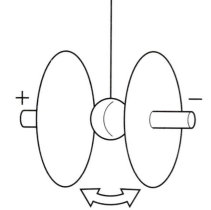

Both the motion of a charge in an electric field, and the forces between charges can be shown by mounting two metal plates vertically and some ten centimetres apart. Then suspend a table tennis ball which has been coated with conducting paint from a thread so that it hangs centrally between the plates. Apply a potential difference between the plates — something over 2 kV is needed. Pull the thread so that the ball touches one of the plates. The ball will then oscillate between the two charged plates transferring charge from one plate to the other.

A moving charge means a current flow and this can be detected using a spot galvanometer since the actual currents are very small, <1 mA. A TV camera is a great help here to show the movement of the ball to the whole class; alternatively, a shadow of the ball can be projected onto a screen.

This experiment has a simple and effective extension. Use the two vertical plates connected to 5 kV and then drop a small piece of aluminium foil between them; it will oscillate backwards and forwards between them as it falls.

Apparatus required: •Two metal plates with insulating handles •Two retort stands and bosses •Table tennis ball •Thread •EHT supply •Foil.

2   Induction and repulsion in electrostatics

A charged polythene rod is brought up to a metal rod that is fixed horizontally on an insulating stand. Touching the other end is a light ball hanging from a thread. As the charged rod is brought closer the ball will swing away from the metal rod. Bringing up the end of a lead fixed to the negative of an EHT supply to a freely suspended ball will achieve the same result.

Apparatus required: •EHT supply •Polythene rod and duster •Conducting ball on insulating thread •Metal rod.

3   Plastic box smoke separator

Two metal plates are mounted parallel to each other in a transparent plastic sandwich box. Blow some smoke into the box and close the lid. Use a van de Graaff generator to apply a potential across the plates and the smoke will clear immediately since the charged plates attract the small particles of smoke towards them. The effect of different potentials can easily be studied.

Apparatus required: •Plastic box smoke separator •van de Graaff generator or EHT supply •Smoke generator.

4 Drawing pin on the van de Graaff: dust/smoke collector

Fix a drawing pin with its point facing upwards in the bottom of a perspex can on top of the van de Graaff generator and blow some smoke into the can. Switch on the generator and observe what happens. The discharge from the drawing pin point collects the smoke into a column. It is due to the concentration of charge at the point and hence the large electric field there. This demonstration also works well without the can, simply using a pin mounted vertically on the dome.

Apparatus required: •van de Graaff generator •Drawing pin in transparent perspex box •Smoke.

5 Some simple electrostatic demonstrations

(a) Charged balloon sticking to the walls or ceiling.
(b) Taking off a jumper over a nylon blouse or shirt.
(c) Moving around under nylon sheets wearing pyjamas or a nightie.
(d) Rubbing your shoes on a synthetic carpet and then touching an earthed metal pole.

Apparatus required: •Balloon •Duster.

6 The charge on your body

Use a dc amplifier or a picocoulomb meter to measure the charge of the human body. This demonstration is very dependent on what the students are wearing on their feet and what the floor is made of but it provokes some discussion!

Apparatus required: •Picocoulomb meter •Suitable output meter.

7 Bubbles and the van de Graaff

This experiment is a novel way of showing electrostatic induction and also the forces between like charges. Blow some soap bubbles near the dome of a van de Graaff generator. Initially they are attracted to the dome but then spring away in the field if they gain the same charge as the dome.

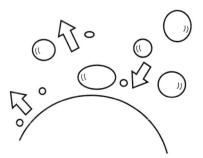

Apparatus required: •van de Graaff generator •Bubble liquid and blower.

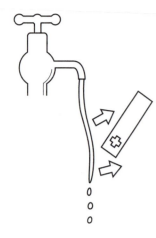

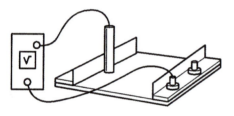

8   A charged rod and its effect on water streams

This next experiment is a beautiful, and to the students quite unexpected, demonstration. Turn on a tap and let a slow but continuous dribble of water fall vertically from it. Bring up a charged polythene rod close to this dribble and the water will be deflected towards it. The deflection is due to the polar molecules (positive at one end and negative at the other) which orient themselves so that they are attracted by the charged rod. A charged plastic comb will also work well and showing this with a TV camera is a help to make it visible to a large group.

Apparatus required: •Polythene rod •Duster •Tap with fine outlet.

9   Electric fields

The apparatus shown in the diagram was home made and has proved very useful for plotting electrostatic fields. A piece of conducting paper is placed over a piece of carbon paper on top of a sheet of white paper and is held in place by the two metal bars. A potential difference is placed across the two bars and using a probe made from an optical pin connected to a digital meter the shape of the field can be plotted. You can also trace the lines of equal potential. Screwing down the bolts in the centre of one or both of the bars enables the fields produced by point charges to be studied.

Apparatus required: •Electric fields apparatus •Digital high resistance voltmeter •Optical pin •Conducting paper •Carbon paper.

10   Spark plug and the van de Graaff

Use a van de Graaff generator to demonstrate the action of a spark plug. Attach the top of the spark plug to the top of the generator using a lead and clip and earth the other contact. The spark plug will operate when the generator is turned on.

Apparatus required: •van de Graaff generator •Spark plug.

11   Electric fields: the school dinner experiment

(a) My rather unkind name for the electric field apparatus. You can make electric fields visible on an overhead projector by using a shallow clear plastic dish of oil into which two electrodes are placed. Sprinkling semolina or fine grass seeds on the surface will show the field lines when the EHT supply (usually 5 kV) is connected between the electrodes and switched on.

 (Safety: EHT.)

(b) Use the classic electric field apparatus on the overhead projector but introduce a metal ring between the electrodes to demonstrate that there is no field inside a hollow conductor.

Apparatus required: •Electric field apparatus •Overhead projector •EHT supply •Oil.

12 Spark counter and ions from a flame

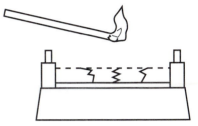

Set up a spark counter with a potential difference of 4 to 5 kV between the wire and the gauze so as to give a voltage not quite sufficient for sparking to take place. To demonstrate that a flame ionizes the air light a taper and blow the flame towards the gauze — sparking immediately results.

Apparatus required: •Spark counter •Taper •EHT supply.

13 Electrostatic fields

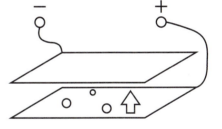

A very good demonstration of electric fields uses the macro-Millikan apparatus, a pair of aluminium plates some 30 cm square and about 5 cm apart which are used as a large scale demonstration of Millikan's oil drop experiment. I have used this with little aluminium foil figures to show electric fields. An even better demonstration is to use figures made from tissue paper, they do not have the problem of being conducting and will rise and fall in the field if it is adjusted carefully. The special apparatus does not have to be used; two aluminium sheets with a potential difference of up to 5 kV between them works perfectly.

(Safety: EHT.)

This experiment will also work if pieces of puffed wheat are used in place of the pieces of paper in the space between two charged plates.

Yet another alternative is to sprinkle a small amount of sugar (or flour) onto the lower plate. Connect the plates to an EHT supply and switch on. The sugar oscillates between the plates showing induced charge, then attraction and then discharge. Some practice will be needed to get the correct potential difference and plate separation, i.e. the electric field intensity.

Apparatus required: •Paper •Sugar •Puffed wheat •Flour •Two metal plates mounted horizontally and separated by polythene spacers •EHT supply.

14   Watch glass electrostatic repulsion

The forces between two charged rods can be shown by the following simple experiment. Balance a charged polythene rod on an inverted watch glass. Then bring up another charged rod towards it. Repulsion and attraction (this time using a cellulose acetate rod) are easily shown.

Apparatus required: •Watch glass •Two polythene rods •Cellulose acetate rod •Duster.

15   Electrostatic forces: balls and a top pan balance

Measure the electrostatic force between two charged balls (or rods), one fixed to a clamp, the other to a top pan balance. Mount the retort stand on a laboratory jack so that the separation of the balls may be altered.

Apparatus required: •Top pan balance reading to at least 0.01 g •Charged rods or balls •Laboratory jack •Retort stand.

16   van de Graaff and a radio

Use a radio to detect the sparks from a van de Graaff generator by the electromagnetic pulse of radiation that they produce. Refer to the effects of lightning on radio and TV transmission. I have found a good response in the medium wave band at a frequency of around 100 kHz.

17   Soap bubbles and Millikan

Set up a van de Graaff generator with the high voltage dome connected to a metal tube held in an insulating clamp. To the lower end of this attach a length of rubber tubing. Cover the top of the tube with a soap film and blow a soap bubble. Mount two metal plates horizontally and connect them either to an EHT supply or to the van de Graaff and earth. Switch on the generator so that the bubble becomes charged. Switch off the generator. Now gently blow the bubble sideways into the gap between the charged plates. By adjusting the plates separation or the potential difference between them (if connected to the EHT), or both, it should be possible to get the bubble to rise, fall or remain suspended in mid air — a useful demonstration of the Millikan experiment. In fact if the mass of the bubble is known we could actually measure the charge on it!

Apparatus required: •Two metal plates mounted horizontally and separated by polythene spacers •Soap bubble liquid •Metal tube •Rubber tubing •van de Graaff or EHT supply.

18 Electrostatic repulsion: various versions

(a) Electrostatic repulsion can be shown by two strips of aluminium foil hanging from the same point and with the join connected to an EHT supply (or a van de Graaff generator). They repel each other as the potential is increased and they become more highly charged. (Safety – remember the high voltages involved!)

(b) Two charged plastic drinking straws are suspended with thread through them so that they hang horizontal and parallel. If they have opposite charge they will swing apart — a very simple demonstration of electrostatic repulsion. As an extension the forces between them can be worked out knowing the masses of the straws and the angle of the threads. A very approximate idea of the charges carried can then be found.

Apparatus required: •Two strips of aluminium foil •EHT supply •Two drinking straws •Thread •Wooden or plastic clamp in stand.

19 Fluorescent tube and the plasma globe

This is a most impressive demonstration. Take an ordinary fluorescent tube and hold it near the plasma globe. This is a globe filled with low-pressure gas that has a ball electrode at a potential of some 20 kV at its centre. Switch on the globe. To most people's amazement the tube lights even though it is not touching the globe. I have found that if one end of the fluorescent tube is a little less than 10 cm from the globe, but not touching it, the demonstration works well. Moving your hand along the tube will 'wipe off' the discharge since you are earthing the tube at that point. It shows the existence of an electric field round the globe, the potential difference between the two ends of the tube being sufficient to make it light.

Apparatus required: •Plasma globe •Fluorescent tube.

20 Statics and covering film

The effects of static electricity can be demonstrated with clingfilm, book covering film or plastic record sleeves. Simply pulling the film off its backing paper will charge both the paper and the film. Testing the two parts with a gold leaf electroscope will show that these are oppositely charged. This can be an annoying property of some types of sticky tape.

Apparatus required: •Roll of book covering film •Clingfilm •Record in plastic sleeve.

21 Photocopier and electrostatics

This is a very simple demonstration of the behaviour of a photocopier. 'Draw' a T shape on a polythene tile by rubbing along the shape with a duster to charge only that area. Now put the tile over a plate of semolina. The semolina will only be attracted to the charged T shape in exactly the same way as carbon powder is attracted to the charged parts of the master drum in a photocopier. Use a coloured tile to make the semolina show up better or use coloured spice powder on a white tile.

Apparatus required: •Semolina •Polythene tile •Duster.

22 Ions in a flame and a flame probe for detection of electrostatic fields

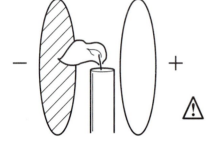

(a) Set up two vertical plates and put a candle between them. Connect the plates to an EHT supply and a spot galvanometer. When the supply is turned on the reading of the meter stays at zero — no current flows. However, when you light the candle, two things happen. The reading of the galvanometer shoots up and the flame is dragged towards one of the plates. The current is due to the ionization of the air between the plates. The ions in the flame itself make it distort. A TV camera pointed between the plates helps to make this visible to the whole class. If one is not available a light source such as a projector can be used to give a shadow of the flame on a screen.
(b) Alternatively a flame from a fine glass jet can be moved within the field. The deflection of the flame due to the movement of the ions within it shows the intensity of the field at that point.

Apparatus required: •Two metal plates mounted on insulating holders •EHT supply •Candle •Glass flame jet.

23 Point discharge

(a) Show the effects of charge discharge from a point using a sensitive flame. The flame is distorted by the flow of charge from a point at a high potential.
(b) The discharge from points can also be easily shown with the rotating windmill apparatus, either mounted on the top of the large dome itself or on a stand. It helps considerably if the stand is placed on a polythene tile to prevent leakage.

Apparatus required: •Windmill of metal with pointed ends •EHT or van de Graaff generator.

24 Pith ball on thread on van de Graaff

Fix a pith ball to a thread and then stick the other end of the thread to the top of the large dome of a van de Graaff generator. When the generator is switched on the ball and dome acquire the same charge and the repulsion between the ball and the dome will make the thread stand up vertically.

Apparatus required: •van de Graaf generator •Pith ball •Thread •Sticky tape.

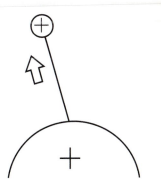

25 Paper and the van de Graaff

A very simple example of electrostatic repulsion is to tear up a sheet of paper into small pieces and put them on top on the large dome of a van de Graaff generator. When the generator is turned on the pieces of paper will fly off — they have all acquired the same charge as the dome and each other and so will be repelled. On a good dry day they can fly almost a metre from the dome! The maximum mass of a piece of paper that is ejected from the dome can be used to estimate the potential of the dome.

Apparatus required: •van de Graaff generator •Paper.

26 The phantom leg

Take a fine weave stocking and wedge a cardboard ring (cut from a storage tube) into the neck to keep it open. Charge the stocking by rubbing it with a balloon until the stocking expands as though it were filled by a 'phantom leg'. This is due to the like charges on all parts of the stocking repelling each other so forcing the sides of the stocking apart.

Apparatus required: •Stocking, fine texture, low denier •Balloon.

27 Field in a hollow charged conductor

Theory predicts that there is no field inside a hollow charged conductor. This is easily demonstrated by taking a gold leaf electroscope, standing it on an insulating tile and connecting the cap to the case. Then charge the top plate. You will notice that the leaf does not rise; all the charge resides on the outside of the electroscope (plate and body) and there is no field within the case.

Theory: electric field E is the negative of the potential gradient ($-\mathrm{d}V/\mathrm{d}x$). So if V is constant $E = 0$.

Apparatus required: •Gold leaf electroscope •Method of charging (charged rod or EHT supply) •Insulating tile •TV camera to make the experiment visible to a large group.

28   Cars and electrostatics

You may have noticed that you may get a shock when getting out of a
car. Here are some (not too serious!) suggestions as to how to prevent this
happening: driving in wellingtons and a wet suit; throwing out an anchor
before you get out; carrying a passenger to jump out first; keeping the
radio aerial up!

CAPACITORS

General theory for this section

Capacitance $= Q/V$; energy stored in a capacitor of capacitance C charged to a potential $V = \frac{1}{2}(QV) = \frac{1}{2}(CV^2)$.

1. Capacitor and bucket of water
2. Timing with a capacitor
3. Capacitor in a camera flash unit
4. The smoothing action of a capacitor

---

1 Capacitor and bucket of water

This is a useful comparison between a charged capacitor and a bucket full of water. A hole at the bottom allows water to run out and represents the current in the circuit as the capacitor discharges – the size of the hole representing the resistance of the circuit.

The depth of water represents the potential difference V across the capacitor and the volume of the bucket represents the maximum charge Q that can be stored by the capacitor when fully charged. You can see that in the same way that a certain volume of water can be stored in either a low but wide bucket or a tall but thin bucket a given amount of charge may be stored at high or low potential in two different size capacitors.

Theory: $Q = CV$ so large V, small C, or small V, large C.

---

2 Timing with a capacitor

Connect a capacitor and a resistor in parallel with the DC supply. Switch on the dc supply and current will flow through the resistor continually but also charge the capacitor. In the supply circuit and the resistor circuit are two thin strips of aluminium foil. When the one in the capacitor circuit is broken the capacitor stops charging and when the one in the resistor branch is broken the capacitor stops discharging. Therefore the capacitor only discharges in the time between the breaking of the two strips. If this time is short then the current from the capacitor is roughly constant and so $Q = It = C\Delta E$ where ΔE is the drop in potential difference across the capacitor. But $E = IR$ and so $t = C\Delta E.R/E$ (useful for bullet timing, falling objects as long as t is small). The pupils could check the validity of this. The potential difference across the capacitor should be measured with a digital voltmeter with a resistance of some $M\Omega$.

Apparatus required: •Capacitor •Aluminium foil •dc supply •Resistor •Stop clock •Digital voltmeter.

3 Capacitor in a camera flash unit

Refer to this as a practical use of capacitors. The energy stored in the capacitor due to the small charging current from the battery over a few tens of seconds is released as one large burst in a fraction of a second.

4 The smoothing action of a capacitor

A capacitor placed across the output of a rectifier unit will give a smoothing effect to the output. An analogy of this is to use a balloon to smooth the flow of water. Fix a tube to a water tap and into this fit a T-piece with a balloon hanging down from one branch of the T. As water flows through the balloon will distend. As the flow decreases the balloon will shrink, maintaining the output flow, and if the flow increases, the balloon expands, smoothing out the flow and keeping the flow rate approximately constant. The expansion and contraction of the balloon is analogous to the charging and discharging of the capacitor.

Apparatus required: •Tap •Tubing •T-piece •Balloon.

ELECTRON PHYSICS

General theory for this section

All materials consist of atoms. These atoms contain shells of orbiting electrons. In metals some of these electrons become detached from the atoms and 'wander' through the material at high speed – they are known as free electrons. There are some 10^{28} of these free electrons per cubic metre in a metal such as copper, but only 10^{22} in a semiconductor at room temperature. The charge on one electron is -1.6×10^{-19} C.

1 Car parks and energy levels

The behaviour of electrons and holes in a semiconductor can be represented by considering cars in a multi-storey car park. The cars represent the electrons and the empty parking spaces the positive holes. Cars can move between parking levels in the car park if there are empty spaces in the same way as electrons move between energy levels if there are empty holes. If the car park is nearly full of cars it is nearly empty as far as the holes are concerned — holes can therefore move freely within it! This only really works as an analogy if all the cars are identical, such as a car park full of minis. The inability of the drivers to decide where to park is analogous to the random motion of the electrons!

2 Atom model

A set of polystyrene balls mounted on wires can be used to show the arrangement of electrons in atoms and the transitions between energy levels. Make a rectangular wooden framework and fix horizontal wires at the correct spacing to represent the energy levels. The polystyrene balls are threaded on vertical wires and can be made to slide along these wires to simulate transitions between one energy level and another.

Apparatus required: •Atom model as described.

3 Hills and pits in the photoelectric effect

The fact that free electrons are held within a metal in a sort of energy pit or potential well can be compared to people in a glass-sided hole — they cannot get out unless they jump out in one go. They cannot go half-way and hang in mid-air waiting for another burst of energy. This is similar to the idea of the quantum of radiation liberating free electrons in photoelectric emission. The electrons have to climb a potential hill to reach a collector and those with insufficient energy cannot do so. Therefore photoelectrons emitted by radiation with a low frequency will be unable to reach a collector if a negative potential is applied to it while those emitted by radiation of a higher frequency will be able to do so. While at London University I took part in a sponsored walk from London to Brighton. Arriving in the outskirts of Brighton I found that the finish was up a hill into a park. Like the electrons I just about had enough energy left for that final climb.

4 Photoelectric effect

In the photoelectric effect quanta of light fall on a surface and liberate free electrons from the surface. This liberation is instantaneous if the frequency of the incoming radiation is high enough.

The photoelectric effect, the quantum nature of radiation and the energies of quanta of different types of radiation can all be demonstrated using ping-pong balls, a billiard ball and a coconut shy. Professor Russell Stannard suggested the original version of this experiment on a splendid old Open University video. A small boy tries to dislodge a coconut by throwing a ping-pong ball at it, with no luck since the ping-pong ball has too little energy! He then tries a whole bowl of ping-pong balls, but the coconut still stays put! Along comes the professor with a pistol and fires one bullet at the coconut — it is instantaneously knocked off its support. This simulates the effect of infrared and ultraviolet radiation on a metal surface. The ping-pong balls represent low-energy infrared, while the bullet takes the place of high-energy ultraviolet. I suggest using a billiard ball in the lab these days rather than a bullet.

If you can get hold of a copy of the original video do watch it carefully; they apparently had a lot of trouble setting it up, and see if you can spot where their particular simulation breaks down!

5 Photoelectric effect – zinc plate

Fix a cleaned (vital to remove oxide) zinc plate to a gold leaf electroscope. Charge it using an EHT supply by touching the plate briefly with the positive or negative lead to show the effects of the sign of the original charge. Then try to discharge it with light from a light bulb, a laser (careful) and a uv lamp (careful). No discharge occurs if the plate was originally positive; if any electrons are emitted they are immediately attracted back to the surface. With an initial negative charge only the high-energy ultraviolet quanta will discharge the electroscope, causing the leaf to fall. Refer to the TV camera tube which is sensitive to visible light.

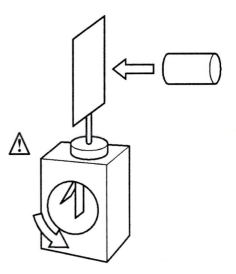

Apparatus required: •Gold leaf electroscope •TV camera if possible •Zinc plate attachment •Ultraviolet light •Laser •Light bulb.

6 Thermionic emission

A hot resistance wire held near the cap of a charged electroscope will cause it to discharge whether it is charged positively or negatively. This is due to the ionization of the air around it by the electrons emitted from the wire. Mount a piece of resistance wire horizontally between two clamps so that it passes about a millimetre over the cap of a gold leaf electroscope. Charge the electroscope. Now pass a large current (around 5 A) through the wire being careful to tension the wire as you increase the current so that it does not sag onto the electroscope cap. The electroscope will discharge.

Apparatus required: •Gold leaf electroscope and means of charging it (polythene rod and duster, EHT or van de Graaff) •Length of resistance wire •Power supply •Two retort stands and clamps.

7 Diode and the depletion layer

I use the idea of a mass of girls and boys in a playground to represent a piece of n-type semiconductor joined to a piece of p-type — in other words a p–n junction. Initially the children are tightly packed together with all the boys in one half of the playground and all the girls in the other. Interesting things only happen at the join, or close to it. A movement of children will take place, eventually preventing any further

216

MODERN PHYSICS

movement. Children far from the join are not affected. This is an analogy for the movement of holes and electrons near the join between two pieces of p- and n-type semiconductor.

8 Fish and thermionic emission

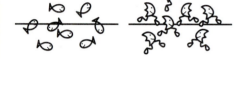

Use the idea of fishpond full of swimming fish as an analogy to explain what is happening inside the cathode of a thermionic valve. Some of the fish will jump out of the water only to fall back; there is a continual jumping out and falling back — a true dynamic equilibrium. This can be compared with the random motion of electrons within a hot metal; some of them gain enough energy to leave the metal surface briefly.

All the fish are the same so you can't tell which fish are in the water and which are in the air. In just the same way free electrons in a metal can give a dynamic equilibrium when the metal is heated — they are continually leaving and re-entering the surface. The attraction of the anode could be represented by a worm lowered in and the heating of the cathode could be compared with a shark in the water making the fishes swim around more vigorously!

9 Nuclear collisions: watches

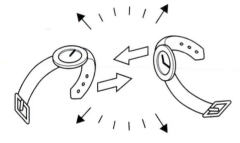

To discover the composition of subatomic particles two particles are fired towards each other at high speed. An example of this is the Large Electron–Positron (LEP) collider at CERN in Geneva, where electrons and positrons are made to collide head on in the giant accelerator. In a way this is rather a crude method. It is a bit like trying to find out what is in a watch by throwing two watches together and watching what bits fly out. The big difference between the analogy and the sub-atomic particles is that in the nuclear collisions some bits that fly out were not there before the collision — they have been 'made' by the conversion of energy to matter in the collision — certainly not true with the watches!

10 Maltese cross

An interesting extension of the traditional Maltese cross tube is to demonstrate the tube but do not connect the cross to the EHT supply. A much fuzzier image of the cross will be formed on the screen. Electrons emitted by the cathode collide with and stick to the cross. This gives the cross a negative charge which repels the electron beam so giving the shadow of the cross a rather distorted and lumpy appearance. This makes a good demonstration of charge repulsion.

Apparatus required: •Maltese cross tube in holder •EHT power supply •Low-voltage power supply.

11 Electron waves in orbits

From Schrödinger's wave equation we can imagine the electron as existing as a wave that fits round an orbit a given number of times. In other words $2\pi r = n\lambda$ where λ is the electron wavelength, n is an integer and r is the radius of the orbit. A simulation of this idea is as follows. A ring of stiff wire should be fitted on top of a vibration generator; I use a loop about 15–20 cm in diameter. By adjusting the frequency of oscillation of the vibration generator, standing waves can be produced in the wire ring analogous to electron waves in the orbits of atoms. Multiple rings are impressive.

An extension of this experiment shows the differing wavelengths further out from the nucleus. Two pieces of cord, one light and one heavy, are tied together. One end is fixed to a retort stand and the other is attached to a vibration generator. When you switch on the generator standing waves are produced in the cords. The frequency is the same in both but the wavelength differs since the speed of the waves in the two cords is different

Apparatus required: •Vibration generator •Loop of stiff wire •Signal generator •Retort stands •Pieces of cord.

12 Energy levels and the escalator

The top steps of an escalator are rather like the energy levels in a hydrogen atom, they get closer together just as the energy levels get closer together as you get nearer the ionization level. This means that as you ascend the escalator you make smaller and smaller energy transitions between successive steps. The analogy is really only good if the escalator is stationary.

13 Accelerator model

Here are two ideas to simulate the action of a particle accelerator where the particle gets a kick every time it passes between two accelerating electrodes, the first a synchrotron (constant radius) and the second a cyclotron.

(a) Make a large circular orbit out of plastic curtain rail. Put a marble inside the ring touching the wall to represent the particle and use an air blower to simulate the kick given by the electric field as the particle orbits the accelerator. If a small section of the track could be made to open you can simulate the extraction of a charged particle from an accelerator by a magnetic field. A ball fixed to a thread might do as well (no accelerator walls). This simulates the operation of a synchrotron where the orbit radius is fixed and the confining magnetic field is made to increase as the particle's energy increases.

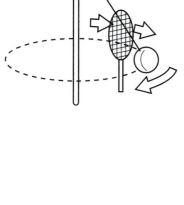

(b) In the second method all you need is a ball suspended by a string from a post (sometimes used to train young children to play tennis) and a bat. Give the ball a push so that it swings round the post and every time it passes you give it a hit with the bat. The ball will gain energy on each orbit, move faster and swinging out further from the post. This simulates the operation of a cyclotron where the magnetic field is constant and the orbit radius increases as the particle's energy increases. Adding a horizontal string to the swing ball fixed between the ball and the vertical central post simulates synchrotron operation. As the ball is hit its velocity increases but the horizontal string prevents it from changing its orbit radius. However, as it gets faster the centripetal force in the string increases — analogous to the increase in magnetic field in a synchrotron required to keep the particles in a constant orbit as they accelerate.

Apparatus required: (a) •Length of plastic curtain rail (at least 2 m) •Air blower •Polystyrene balls or marbles. (b) •Ball on string tied to the top of a fixed post •Bat.

14 The bubble chamber

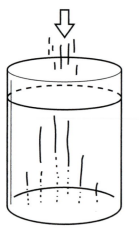

This simple experiment is an excellent demonstration of how a bubble chamber works and can also be used to reinforce the idea of a cloud chamber. Pour out a gassy drink such as lemonade and allow it to stand for a few minutes until no more bubbles rise. Then drop in a small amount of fine salt (or sand). This gives nuclei for bubbles to form on and you will see clouds of bubbles begin to rise through the liquid. In a similar manner ions are formed in a bubble chamber when a radioactive particle passes through and bubbles form on these. Bubble chambers are more effective than cloud chambers as there are more atoms that can be ionized and hence more 'nuclei' on which bubbles can form.

Apparatus required: •Can of fizzy drink •Glass beaker •Salt.

NUCLEAR PHYSICS

General theory for this section

The radius of a nucleus is given by the equation $r = r_o A^{1/3}$, where A is the nucleon number of the nucleus and r_o is a constant (1.3×10^{-15} m)

1 Simple atom analogy/model

A simple way of remembering the relative sizes of the nucleus and the atom of hydrogen is to imagine that if the nucleus is represented by a ball 1 cm in diameter (say a marble) then the orbiting electron would be 1 km away. The ratio of the diameter of the atom to that of the nucleus is about 1 to 100 000.

2 Plum pudding model

If we are going to use the special metal tin hat model to simulate the nuclear model of an atom then we should show the alternative plum pudding simulation. Instead of the 'tin hat' I have used a cymbal – the ball bearing will roll across the centre quite easily showing reduced scattering by this type of atom. There is a deflection but if the ball bearings are rolled from about halfway up the little ramp no ball bearing ever suffers more than a 90° deflection.

Theory: the much lower electric field intensity of the plum pudding model is represented by the lower central height of the cymbal. The potential at the centre is also much smaller resulting in a smaller deflection by the incident alpha particles.

Apparatus required: •Cymbal •Ball bearings.

3 Alpha particle scattering: two alternatives

There are two simple alternatives to the commercially available 'tin hat' apparatus to demonstrate alpha particle scattering by a nucleus.

(a) Fix a rubber sucker to a glass sheet placed on an overhead projector. Now roll a ball bearing towards it. The disadvantage of this method is the relatively small size of the sucker.

(b) Use a fixed circular magnet representing the nucleus and then swing another small magnet towards it to represent the alpha particle (in this version it is important to keep a repulsion between the two magnets by using a suitable method of suspension)

Apparatus required: •Overhead projector •Ball bearings •Rubber sucker •Magnets.

4 Rutherford scattering: a permanent record

A permanent record of the Rutherford scattering experiments can be obtained by using the tin hat simulation of the field but putting it on a piece of white paper resting on carbon paper. Draw a set of lines on the white paper to show the direction of incident ball bearings. Roll the ball bearings down a ramp and the pressure on the carbon paper will give a permanent record of their tracks after 'collision' with the 'nucleus'. (The ball bearings can be collected by surrounding the apparatus by a magnetic strip or simply just catching them!)

Apparatus required: •Tin hat apparatus •Ball bearings •White paper •Carbon paper •Flexible magnetic strip (optional).

5 Mousetraps and chain reactions

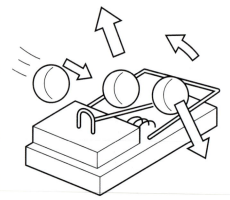

This lovely (although potentially painful) demonstration is a splendid simulation of a nuclear chain reaction. Set up a number of cocked mousetraps side by side in a rectangle (a dozen will work well if you can manage it!) to represent uranium nuclei. Put a couple of polystyrene balls representing neutrons on each one.

Throw in another polystyrene ball (neutron) to start the chain reaction. One mousetrap goes off, this sets off others and so on!

The balls are shot all over the lab (fast neutrons). Spacing out the mousetraps gives an idea of shape of fuel rods and the way a chain reaction can fail. It works best if the mousetraps are placed on the lid of a cardboard box to give a good 'linkage' between one 'fission' and another.

Alternative methods are to use a set of matches fixed upright in a block of wood: when you light one the others are set alight as the flame spreads through them like a forest fire. You could also knock down a set of dominoes by touching the first one and making it fall so that it collides with another and so on.

Apparatus required: •Mousetraps •Polystyrene balls •Block of wood with holes •Matches •Dominoes.

---

6   £20 note and ball: nuclear forces

These two analogies are designed to explain the forces between sub-nuclear particles.

(a) A simulation of long range electrostatic repulsive forces can be given by two people throwing a ball to each other. Each one stands on a skateboard or wheeled trolley. The thrower recoils when they throw the ball and the catcher recoils when they receive it. There is no limit to the range of the force as long as they throw hard enough.

(b) The short range strong nuclear force (only effective over distances less than about 10^{-15} m) can be represented by the two people trying to throw a £20 note to each other. For a start there is no possibility of them throwing it to each other unless they are close together because of the large amount of air resistance; it will only work at short distances but it is certainly attractive!

Theory: both these experiments are simulations of the exchange of force carriers when a force acts between two objects — a photon in the case of the electromagnetic force and a gluon in the case of the strong nuclear force.

Apparatus required: •Two heavy balls •Two skateboards •Two notes.

---

7   Rutherford scattering

Use a row of pins hidden inside a box. Roll ball bearings into one side of the box and observe where they come out. Get students to predict what sort of obstacles are in the box.

---

8   Nuclear fusion analogy

The problem of the repulsion between two nuclei and the existence of a short-range nuclear force can be explained as follows. Imagine two people wearing large inflatable suits so that they appear to be standing in the centre of a balloon. The suits are so large that if they stand near each other they cannot hold hands. Only by the people running together at high speed will the suits be sufficiently squashed for them to be able to grasp each other's hand and so 'fuse'!

QUANTUM PHYSICS

1. Flight of stairs and the quantum theory
2. Quantum theory and milk
3. When the quantum theory becomes effective

1   Flight of stairs and the quantum theory

One of the basic results of the quantum theory is that some energy states are not allowed. Because of the very small value of Planck's constant (6.6 $\times$ 10^{-34} J s) we do not usually observe this discontinuity. However we could compare it to a flight of stairs down which a man and an ant are travelling. The steps are effectively classical to the man — he seems to travel down with an uninterrupted energy change, but they are of a quantum nature to the ant. It all depends on your scale of observation.

2   Quantum theory and milk

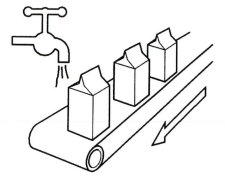

A man wants exactly a pint of milk. He can get it either by filling a bottle from a tap or by catching a carton as it comes off a conveyor belt. Filling the bottle from the tap represents the classical nature of physics while the cartons represents the quantum idea. Notice that with the quantum analogy he may get a quantum (carton) immediately the belt is switched on or have wait for some time for a carton to arrive. However, when it does he will still get the whole carton at once. With the classical theory he gets milk as soon as the tap is turned on but the rate of flow may be small and he may not get a pint as quickly as he does by the quantum method!

3   When the quantum theory becomes effective

This analogy attempts to show when the quantum theory would be observable in every day life, a consequence of the value of Planck's constant being very large, i.e. close to one. Imagine a soccer player trying to score a goal through a slot in a defensive wall. In normal life he would expect to be able to hit any part of the goal simply by changing his shooting position. However if he finds that there are some places the ball cannot get to no matter where he stands then that's when he has entered the realm of quantum soccer!

RADIOACTIVE DECAY AND HALF-LIFE

General theory for this section

The decay of a radioactive source is a random process. The half-life of a source is the time taken for the activity of that source to decrease by half. Remember that the half life of a source is independent of any other physical process such as temperature, pressure, velocity, magnetic and electric fields, etc.

1. Simple radioactive decay formula
2. Alpha radiation: tissue paper and cling film
3. Half-life of water
4. Half-life: wooden or plastic blocks
5. The radioactive decay series

1 Simple radioactive decay formula

This was suggested by one of my sixth form students and seems a simple alternative to the normal decay formula ($N = N_o e^{-\lambda t}$ or $A = A_o e^{-\lambda t}$) (A_o is the original activity of the source, A the activity after a time t and λ the disintegration constant ($\lambda = \ln2/T$ where T is the half-life)). For very simple decay problems we use the fact that the activity will decrease by a factor of 2 in one half-life, 4 in two half-lives, 8 in three half-lives, and so on. Why does the number of half-lives have to be a whole number (of course it doesn't) so lets call the number of half-lives that have passed n, where n is any number. The decay formula suggested is then $A = A_o/2^n$. It always works!

2 Alpha radiation: tissue paper and cling film

The absorption of alpha radiation can best be demonstrated with tissue paper, such as one sheet of a double paper handkerchief. Ordinary paper is rather thick and will stop much of the radiation. It is also worth trying it with very thin aluminium leaf! The alpha particles should go through. A piece of bacon is the closest thing that we can get to show the absorption of radiation by human flesh.

Another material that can be used to test the penetration of alpha particles is a sample of clingfilm. The particles will pass through it.

Apparatus required: •Tissue paper •Aluminium leaf •Geiger counter and display •Alpha source in holder •Clingfilm in holder.

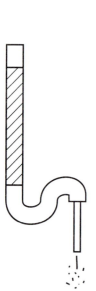

3   Half-life of water

The drop in level of water in a glass tube can be used as an analogy of radioactive decay. In just the same way as the activity of a source decreases as the number of radioactive nuclei decreases so the volume of water coming from the outlet at the bottom of the tube decreases as the height of water in the tube gets less.

Use a vertical glass tube with a short length of capillary tubing fixed to the bottom of it using a length of rubber tubing. Close the rubber tube with a tube clip and fill the large tube with water. Open the tube clip and allow the water to flow out through the capillary tube. Measure the height of the water against time and plot a graph of height against time. It gives a very good analogy of radioactive decay!

Theory: $N = N_{o}e^{-\lambda t}$ or $h = h_{o}e^{-Ct}$

Apparatus required: •Retort stand •Two bosses and two clamps •Stopwatch •Ruler •Glass tube •Rubber tube •Capillary tube •Tube clip.

4   Half-life: wooden or plastic blocks

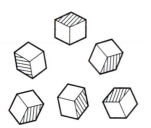

This is a classic experiment that gives a very good analogy with radioactive decay. A large number (1000) small wood (or plastic) blocks (side 1 cm) are given out to the class. Each block has one face coloured or marked with a dot (a set of dice would do but are rather expensive). Each group records the initial number of blocks and then throws them, taking out any that fall with the coloured side up. They then throw the ones left, i.e. those that fell with a plain side uppermost and then repeat the process, until they have none left (usually about 20 throws).

The data from the whole class are collected and a graph plotted of the number of blocks remaining after each throw. It helps to use a spreadsheet for the collection of results if a computer is available in the lab. The idea of the random way in which the blocks fall should be compared with the random nature of radioactive decay.

I have extended this experiment to my older students. In this extension I use the spreadsheet to process my results and get a print out of number of blocks (radioactive nuclei) (N) against throw (time). It is worth doing this after just one or two groups had thrown the blocks showing that as more and more blocks are added to the experiment the line becomes smoother.

Then draw tangents to the curve at know values of N and plot the rate of decay (dN/dt) against N; my class obtained a perfect straight line. This makes a very useful point for a discussion of the equation $dN/dt = -\lambda N$.

Then the computer works out ln (N) and plots this against t — a perfect straight line with a negative gradient.

Apparatus required: •Six-sided cubical wood blocks, one side coloured •Computer with spreadsheet •Printer.

---

5 The radioactive decay series

Having been concerned with the mathematics of a source producing a daughter product which then decays itself into a stable isotope, I decided to modify the previous two experiments to demonstrate this. We have to adapt experiment 4 by using six-sided dice for the initial radioactive material and then ones with a greater number of faces (say ten) for the daughter product.

In the half-life of water experiment the water from the first tube runs into a second tube with a narrower outlet.

Theory: For element B, $dN_B/dt = \lambda A N_A - \lambda_B N_B$

$dN_B/dt + \lambda_B N_B = \lambda_A N_0 e^{-[\lambda t]}$

Solving this gives:

$N_B = N_0[ae^{-[\lambda t]} - be^{-[\lambda t]}]$ where $a = \lambda_A/(\lambda_B - \lambda_A)$ and $b = -a$.

Apparatus required: •A large number of six-sided and ten-sided dice.

ASTRONOMY

1. Volcanoes and heat energy
2. Rotation period of the Moon
3. The echo of the big bang
4. Crater experiment
5. When the universe was dark! 300 000 to one billion years after the big bang
6. Seasons
7. Expansion of the universe
8. The minimum age of the universe
9. Astronomical distances simplified
10. Some simple experiments in astronomy

1 Volcanoes and heat energy

Although most craters on the Moon and the planets are thought to have been created by meteor impact some will have been formed by volcanic activity. You can use a saucepan full of custard or porridge being heated to demonstrate the 'breakthrough of internal magma'! The air escaping from beneath the surface when they are heated gives a very good simulation of volcanic activity.

Apparatus required: •Porridge or custard •Saucepan •Heater.

2 Rotation period of the Moon

This simple demonstration shows why it is that the Moon always presents the same face towards the Earth. Two students are needed to demonstrate this. One student is used to represent the Earth, the other to represent the Moon. One pupil stands still while the second pupil faces inwards and then moves round the first making one rotation as they orbit the 'Earth' once. You will notice that the 'Moon' always faces the 'Earth'.

3 The echo of the big bang

After the enormous temperatures of the big bang the universe cooled. Fifteen thousand million years later (now) the temperature of the universe is about 3 K. We can detect the radiation produced by this temperature, called the echo of the big bang. It has been estimated that about 1% of the background hiss on your television set is due to the after-effects of this enormous fireball.

4 Crater experiment

This experiment is a simulation of meteor impact forming craters on a planet. The meteors are replaced by ball bearings or wooden balls and a tray of sand replaces the planet surface. (I have used icing sugar, both dry and mixed with water, but it is not so good.) The projectiles are dropped from a known height into the sand and the loss of potential energy is related to the crater size. We measured the diameter, depth, wall height and general shape of the craters and also tried oblique impacts by rolling the ball bearings down a tube. Dampening the sand gives a way of investigating different types of planetary surface.

Apparatus required: •Sand •Flour •Ball bearings •Marbles •Wooden balls •Plasticine •Ruler •Balance.

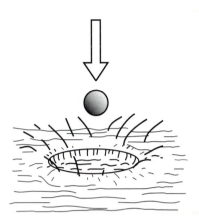

5 When the universe was dark! 300 000 to one billion years after the big bang

As the universe cooled down after the big bang a point was reached where the temperature was such that the emitted radiation was in the infrared – in other words there was no visible light. The universe became dark. This was to remain the state until the first stars fused their first hydrogen and this was to happen some billion years later!

6 Seasons

The seasons occur because of the different area of the Earth's surface over which the sunlight spreads at different times of the year. Use a projector to represent the Sun, and a board that can be rotated representing the surface of the Earth to show this change of area and an LDR to measure the change in light intensity. As the area on which a certain amount of sunlight falls is larger in winter it is therefore colder.

Apparatus required: •Projector •Board in clamp that can be tilted •LDR •Method of measuring the angle of tilt.

7 Expansion of the universe

Here are three possible analogies of the expansion of the universe.

(a) A balloon with dots marked on it to represent the galaxies. As the balloon is blown up all the 'galaxies' recede from each other.

(b) A loaf of bread with currants in it shows a 3D analogy of the recession of the galaxies. The currants represent the galaxies. As the loaf expands when it is cooked all the 'galaxies' recede from each other.

(c) Thread three or four large polystyrene balls onto a length of elastic, one end of which is fixed to a hook on the wall or a secure retort stand. Now pull the free end of the elastic; all the balls separate from each other.

The point about all these analogies is that they make their own space as they expand. Similarly before the universe began to expand there was no space! Actually there was no time either — both space and time were 'created' at the start of the universe.

8 The minimum age of the universe

A minimum age for the universe can be suggested by considering the elements that exist today. In ordinary stars fusion occurs, building up elements as heavy as iron 56 but no further — they are not hot enough. Heavier elements are formed in the enormous explosions of supernovae where the temperatures reached are much greater than those in a normal star. Therefore the universe must be at least old enough for these supernovae to have formed and exploded in order to give us the abundance of heavy elements necessary for the universe to be as we know it today!

9 Astronomical distances simplified

If we represent the distance from the Earth to the Sun as one unit then the scale of the solar system can be simplified as follows:

Sun to:
Mercury 1/3   Venus 2/3   Earth 1   Mars 1.5   Jupiter 5   Saturn 10
Uranus 20   Neptune 30   Pluto 40.

The size of the universe is best considered by expressing distance in terms of the time it would take light to travel that distance at 300 000 km per second:
Earth to the Sun, 8 minutes; across the solar system, 11 hours; to the nearest star, 4 years; across the galaxy, 100 000 years; From our galaxy to the next, 1 million years; To the 'edge' of the observable universe, 15 000 million years.

10 Some simple experiments in astronomy

(a) Diameter of the projected image of the Sun to show the varying distance of the Earth from the Sun during a year.
(b) A sundial.
(c) Star trails using a camera and an exposure of an hour or more.
(d) Star pictures using a camera with an exposure of five minutes.
(e) Sunspots using a projected image of the Sun.
(f) Length of a shadow at noon at two points on the Earth's surface lying

on a north–south line used to measure the size of the Earth (after Eratosthenes).

(g) Observe the paths of meteors during a meteor shower. The shower dates are:

Quarantids, January 1–6; Lyrids, April 19–25; Eta Aquarids, April 24–May 20; Perseids, July 25–August 20; Orionids, October 15–November 2; Taurids, October 15–November 25; Leonids, November 15–20; Geminids, December 7–17.

(h) Plot planetary paths.

(i) Make a Foucault pendulum to show the rotation of the Earth.

(j) Use a telescope with a microscope camera to video the lunar surface; remove both the eyepiece of the telescope and the camera lens.

Warning: never look directly at the Sun — especially through binoculars or a telescope.

Homemade filters should never be used when observing the Sun with binoculars or a telescope.